Miroslav Bačák

Convex Analysis and Optimization in Hadamard Spaces

De Gruyter Series in Nonlinear Analysis and Applications

Volume 22

Preface

Convex analysis and consequently also convex optimization have been traditionally developed in the realm of linear spaces. In this book we attempt to show that a rich theory of convex functions and convex sets can be built also in spaces with no linear structure, namely in Hadamard spaces. Surprisingly, many of the classical results we came to expect in Euclidean or Hilbert spaces extend into this nonlinear setting. On the other hand some areas of convex analysis have so far no counterparts in Hadamard spaces and this offers a challenge for years to come.

The history of Hadamard spaces and their role in geometry and geometric group theory have been described in a number of books, for instance [3, 25, 51, 56, 104], and we refer the interested reader therein. Since convexity is an important element in Hadamard space geometry, convex analysis in these spaces is believed to arise as a rightful subject field. It can be also viewed as a subfield of analysis in metric spaces [8], and then it turns out that a Hadamard space is the metric space where the strongest results can be obtained.

Building upon convex analysis, convex optimization provides us with algorithms for solving a variety of problems which may appear in sciences and engineering. We will describe an application of convex optimization in Hadamard spaces to computational biology in the last chapter.

One of the nice features of the theory presented in this book is that it originated in a pleasant mix of subject areas. The tools and methods are far from being restricted just to the analytical ones. Indeed, we shall see that analysis, geometry, probability, optimization and combinatorics are all mixed together here in a natural and fruitful way, and convexity is an underlying property which stands behind every single result.

Let us now briefly outline the origins of convex analysis in Hadamard spaces. Since 1990s there has been a considerable interest in generalized harmonic maps from a measure space into a Hadamard space, which was triggered by the seminal papers [96, 102, 121] and made a bridge from geometry to analysis in Hadamard spaces. When developing the theory of generalized harmonic maps, J. Jost established numerous convex analytical results in Hadamard spaces. His contributions include resolvents of convex functions, gradient flow semigroups, Γ-limits of convex functions [103, 105] and barycenters of probability measures [102]. At the same time, gradient flows were independently studied by U. Mayer [139], who was the first to consider parabolic problems in Hadamard space. We would also like to mention a related work in a hyperbolic space due to S. Reich and D. Shoikhet [175]. Substantial advancements in the theory of generalized harmonic maps were later achieved by K.-T. Sturm via a probabilistic approach. As a matter of fact, he developed probability theory in Hadamard spaces and also surveyed many already existing results in an extraordinarily lucid way [192, 193, 195, 196]. A number of other authors have contributed since then to convex analysis in Hadamard spaces and they shall be mentioned in the course of our text. Each chapter ends with bibliographical remarks describing where the material comes from.

In this book we intend to present a systematic theory of convex functions, convex sets and mappings from a measure space into a Hadamard space from the rudiments of the theory to the more advanced results which have already achieved a sufficient level of maturity. The only existing book on analysis in Hadamard spaces is by J. Jost [104] and it is our aim here to cover also those parts of the theory which have appeared since it came out. Our exposition is rather self-contained. The only prerequisite which is anticipated is a positive attitude towards nonpositive curvature.

This book was written at the Max Planck Institute in Leipzig. It is my great pleasure to thank Jürgen Jost and Martin Kell for all the discussions on the subject we had during the time. Special thanks go to Simeon Reich for his very valuable comments on the manuscript. Finally, I am grateful to the people at De Gruyter for their very efficient and professional work during the whole process of publication.

The research leading to these results has received funding from the European Research Council under the European Union's Seventh Framework Programme (FP7/2007-2013) / ERC grant agreement no 267087.

We acknowledge the existence of IPE, the drawing editor, which we used to draw all the pictures.

Contents

1 Geometry of nonpositive curvature

The first chapter is devoted to basic notions in metric spaces including a geodesic, metric midpoint and an angle. We define geodesic metric spaces and then explain how one can compare geodesic triangles in these spaces with triangles in the Euclidean plane. Such comparisons enable us to define nonpositive curvature in geodesic spaces and hence to define Hadamard spaces. The condition of nonpositive curvature can be neatly expressed by an analytical inequality and, since we are concerned more with analysis than geometry, this is how we shall use it in our developments. We however feel it is also helpful to gain a geometrical intuition for Hadamard spaces and therefore we start with the triangle comparisons. The chapter ends by providing a number of equivalent conditions for Hadamard spaces.

Unless stated otherwise, the 2-dimensional vector space $\mathbb{R}^2$ is assumed to be equipped with the Euclidean norm

$$\|x\| := \sqrt{x_1^2 + x_2^2}, \quad x = (x_1, x_2) \in \mathbb{R}^2 .$$

The corresponding inner product is denoted $\langle \cdot, \cdot \rangle$.

1.1 Geodesic metric spaces

Let (X, d) be a metric space. A continuous mapping from the interval $[0, 1]$ to X is called a *path*. The *length* of a path $\gamma\colon [0, 1] \to X$ is defined as

$$\operatorname{length}(\gamma) := \sup \sum_{i=1}^{n} d\left(\gamma\left(t_{i-1}\right), \gamma\left(t_i\right)\right) ,$$

where the supremum is taken over the set of all partitions $0 = t_0 < \cdots < t_n = 1$ of the interval $[0, 1]$, with an arbitrary $n \in \mathbb{N}$. Given a pair of points $x, y \in X$, we say that a path $\gamma\colon [0, 1] \to X$ joins x and y if $\gamma(0) = x$ and $\gamma(1) = y$. A metric space (X, d) is a *length space* if for every $x, y \in X$ and $\varepsilon > 0$ there exists a path $\gamma\colon [0, 1] \to X$ joining x and y such that $\operatorname{length}(\gamma) \le d(x, y) + \varepsilon$. If $\gamma\colon [0, 1] \to X$ is a path and $t \in [0, 1]$, we often use the symbol γ_t to denote the point $\gamma(t)$.

A path $\gamma\colon [0, 1] \to X$ is called a *geodesic* if $d(\gamma_s, \gamma_t) = d(\gamma_0, \gamma_1)|s - t|$ for every $s, t \in [0, 1]$, that is, if it parametrized proportionally to the arc length. In particular, a geodesic is an injection unless it is trivial, that is, unless $\gamma_0 = \gamma_1$. When no confusion is likely, we do not distinguish between a geodesic $\gamma\colon [0, 1] \to X$ and its range $\gamma([0, 1]) \subset X$. This allows to say, for instance, that a point *lies* on a geodesic, or a geodesic *goes through* a point.

We say that (X, d) is a *geodesic space* if every two points $x, y \in X$ are connected by a geodesic. In this case we denote such a geodesic by $[x, y]$, but one must remember that such a geodesic is not uniquely determined by its endpoints in general. For a point

$z \in [x, y]$ we often use the notation $z = (1-t)x + ty$, where $t = d(x,z)/d(x,y)$, and say that z is a *convex combination*[1] of x and y. If $x_0, x_1 \in X$ and $t \in [0,1]$, then the point $(1-t)x_0 + tx_1$ on a geodesic $[x_0, x_1]$ is usually denoted $x(t)$ or x_t. The point $x_{\frac{1}{2}}$ is called a *midpoint* of x_0 and x_1. Every geodesic space is a length space; for the converse, see Exercise 1.6.

If moreover every two points of (X,d) are connected by a unique geodesic, the space (X,d) is called *uniquely geodesic*. Since we deal with Hadamard spaces in this book, which are uniquely geodesic, the symbol $[x,y]$ will then denote the unique geodesic connecting x and y, and no confusion can occur. In particular, we have uniqueness of midpoints.

Having defined a midpoint of a given pair of points in a geodesic space, we now introduce a more general notion in an arbitrary metric space.

Definition 1.1.1 (Metric midpoint). Let (X,d) be a metric space and $x, y \in X$. We say that a point $m \in X$ is a *metric midpoint* of x and y if $d(x,y) = 2d(x,m) = 2d(m,y)$, and we say that a pair of points x and y has *approximate metric midpoints* if for every $\varepsilon > 0$ there is $a \in X$ such that

$$\max\left\{d(x,a), d(y,a)\right\} \le \tfrac{1}{2}d(x,y) + \varepsilon .$$

If (X,d) is a geodesic space and $x_0, x_1 \in X$, then a midpoint $x_{\frac{1}{2}} := \frac{1}{2}x_0 + \frac{1}{2}x_1$ is of course also a metric midpoint of x_0 and x_1.

The existence of approximate metric midpoints characterizes length spaces among all complete metric spaces.

Proposition 1.1.2. *Let (X,d) be a complete metric space. Then the following are equivalent:*
(i) *The space (X,d) is a length space.*
(ii) *For every $x, y \in X$ and $\varepsilon > 0$ there exists a point $b \in X$ such that*

$$d(x,b)^2 + d(y,b)^2 \le \tfrac{1}{2}d(x,y)^2 + \varepsilon.$$

(iii) *Every pair of points in X has approximate metric midpoints.*

Proof. Exercise 1.5. □

In a similar way, one can characterize geodesic spaces via the existence of metric midpoints. Indeed, every geodesic space obviously admits metric midpoints. The converse is also true in complete spaces as the following proposition shows.

Proposition 1.1.3. *Let (X,d) be a complete metric space. Then the following are equivalent:*
(i) *The space (X,d) is a geodesic space.*

1 Note that we do not define a convex combination of three or more points.

(ii) *For every* $x, y \in X$ *there exists a point* $m \in X$ *such that*

$$d(x,m)^2 + d(y,m)^2 = \tfrac{1}{2}d(x,y)^2 .$$

(iii) *Every pair of points in* X *has a metric midpoint.*

Proof. Clearly (i) $\Longrightarrow$ (ii), and by Exercise 1.1 also (ii) $\Longrightarrow$ (iii). To see that (iii) $\Longrightarrow$ (i) we choose a pair of points $x_0, x_1 \in X$ and repeatedly employ (ii) to obtain x_t for every dyadic $t := k/2^l$, where $k = 1, \ldots, 2^l$ and $l \in \mathbb{N}$. Since X is complete, we obtain a geodesic $x\colon [0,1] \to X$ connecting x_0 and x_1. □

In order to give the geometrical definition of nonpositive curvature we need to make precise how to compare geodesic triangles with triangles in the Euclidean plane.

Geodesic triangles

Let (X,d) be a geodesic metric space. A *geodesic triangle* with vertices $p, q, r \in X$ consists of three geodesics $[p,q], [q,r], [r,p]$. If the space is not *uniquely* geodesic, a geodesic triangle is not uniquely determined by its vertices. We will however denote it by $\Delta(p,q,r)$, keeping in mind that the definition implicitly presumes a choice of geodesics $[p,q], [q,r], [r,p]$.

Given such a geodesic triangle in X, there exists a *comparison triangle* in $\mathbb{R}^2$, that is, three line segments $[\overline{p},\overline{q}], [\overline{q},\overline{r}]$ and $[\overline{r},\overline{p}]$ in $\mathbb{R}^2$, such that

$$d(p,q) = \|\overline{p} - \overline{q}\|, \quad d(p,r) = \|\overline{p} - \overline{r}\|, \quad d(r,q) = \|\overline{r} - \overline{q}\|,$$

see Figure 1.1. The comparison triangle is unique up to isometries and we shall denote it $\Delta(\overline{p},\overline{q},\overline{r})$. Note that both a geodesic triangle and a comparison triangle are a union of three geodesics, nothing like a convex combination of three points. The orientation of the sides plays no role, meaning that geodesics $[p,q]$ and $[q,p]$ represent the same side of a triangle. Let us now assume $x \in \Delta(p,q,r)$, say $x = (1-t)p + tq$ for some $t \in [0,1]$. Then the *comparison point* for x in the comparison triangle $\Delta(\overline{p},\overline{q},\overline{r})$ is the point $\overline{x} := (1-t)\overline{p} + t\overline{q}$. A word of caution is in order: the triangle $\Delta(\overline{p},\overline{x},\overline{r})$ is *not* a

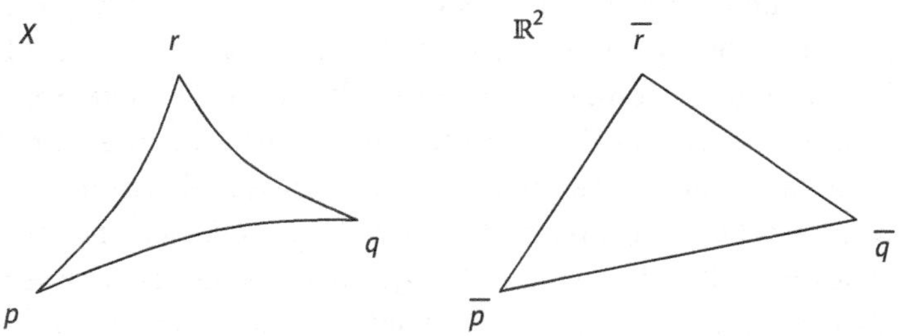

Fig. 1.1. Geodesic triangle (left) and its comparison triangle (right).

comparison triangle for the geodesic triangle $\triangle(p, x, r)$, because in general we do not have $d(x, r) = \|\overline{x} - \overline{r}\|$.

Next we get to the definition of nonpositive curvature in the sense of Busemann. This is however only an intermediate step for us, since we will need a stronger notion of nonpositive curvature, which is to appear later (more precisely in Definition 1.2.1).

Definition 1.1.4 (Busemann nonpositive curvature). Let (X, d) be a geodesic space. We say that X has *nonpositive curvature in the sense of Busemann* if for every $x, y, z \in X$ we have $2d(m_1, m_2) \leq d(x, y)$, where m_1 is a midpoint of $[x, z]$ and m_2 is a midpoint of $[y, z]$. We will alternatively say that (X, d) is a *Busemann space.*

The geometric condition in the definition means that a triangle in a Busemann space is thinner than its comparison triangle in the Euclidean place. It has an analytic counterpart in terms of convexity of the distance function.

Proposition 1.1.5. *A geodesic space (X, d) is a Busemann space if and only if, for every geodesics $x, y\colon [0, 1] \to X$, the function $t \mapsto d(x_t, y_t)$ is convex on $[0, 1]$.*

Proof. It of course suffices to prove the implication that in Busemann spaces the function $t \mapsto d(x_t, y_t)$ is convex. Let m be a midpoint of x_1 and y_0. We then have

$$d(x_{\frac{1}{2}}, y_{\frac{1}{2}}) \leq d(x_{\frac{1}{2}}, m) + d(m, y_{\frac{1}{2}}) \leq \frac{1}{2} d\left(x_0, y_0\right) + \frac{1}{2} d\left(x_1, y_1\right).$$

From this we easily obtain

$$d\left(x_t, y_t\right) \leq (1-t) d\left(x_0, y_0\right) + t d\left(x_1, y_1\right),$$

for every dyadic $t := k/2^l$, with $k = 1, \ldots, 2^l$ and $l \in \mathbb{N}$. An easy approximation argument extends the last inequality for every $t \in [0, 1]$. □

It is immediate that Busemann spaces are uniquely geodesic. A typical example of Busemann spaces is a strictly convex Banach space.

Even though Busemann's definition captures the essence of nonpositive curvature very well indeed (a complete simply connected Riemannian manifold has nonpositive sectional curvature if and only if it is a Busemann space), it is too weak for our purposes. A major disadvantage of Busemann spaces is that metric projections onto closed convex sets are not nonexpansive in general. Since nonexpansiveness of metric projections underpins so many results in the present book, we therefore need a stronger notion of nonpositive curvature due to Alexandrov. Using Gromov's terminology, we will refer to spaces of nonpositive curvature in the sense of Alexandrov as CAT(0) spaces. The difference between the Alexandrov curvature and Busemann curvature is similar to that between Hilbert spaces and strictly convex Banach spaces; see Example 1.2.9. It is worth mentioning that there is also a notion of uniform Busemann spaces, which in a sense correspond to uniformly convex Banach spaces.

Actually, there is a precisely formulated difference between Busemann's and Alexandrov's nonpositive curvatures. A geodesic metric space has nonpositive curvature in the sense of Alexandrov if and only if it has nonpositive curvature in the sense of Busemann and it is a Ptolemaic metric space; see Remark 1.3.4 below.

1.2 Meet Hadamard spaces

We now introduce CAT(0) spaces, which are geodesic spaces with nonpositive curvature in a stronger sense than that of Busemann. Like in Busemann curvature, we compare geodesic triangles of a CAT(0) space with triangles in the Euclidean plane, but require a stronger inequality to hold. Similarly, we will also see that this geometric condition can be expressed analytically as convexity of the distance function and that this convexity condition is quantified. Complete CAT(0) spaces are called Hadamard spaces (Definition 1.2.3) and these are already the spaces we are going to use throughout the book.

Definition 1.2.1 (CAT(0) space). Let (X, d) be a geodesic space. It is a CAT(0) space if for every geodesic triangle with vertices $p, q, r \in X$ and $x \in [p, r]$, $y \in [p, q]$ we have $d(x, y) \leq \|\overline{x} - \overline{y}\|$, where $\overline{x}$ and $\overline{y}$ are the corresponding comparison points in the comparison triangle $\triangle(\overline{p}, \overline{q}, \overline{r})$.

The geometrical meaning of this definition is depicted in Figure 1.2. One can see from the very definition that CAT(0) spaces are Busemann. In particular, they are *uniquely* geodesic. If (X, d) is a CAT(0) space, then every geodesic triangle with vertices $p, q, r \in X$ and $x \in [q, r]$ we have

$$d(p, x) \leq \|\overline{p} - \overline{x}\| , \tag{1.2.1}$$

where $\overline{x} \in \overline{\triangle}(\overline{p}, \overline{q}, \overline{r})$. Using an elementary calculation with the inner product in the Euclidean plane $\mathbb{R}^2$, we can equivalently express the condition in (1.2.1) as

$$d(p, x_t)^2 \leq (1 - t)d(p, q)^2 + td(p, r)^2 - t(1 - t)d(q, r)^2 , \tag{1.2.2}$$

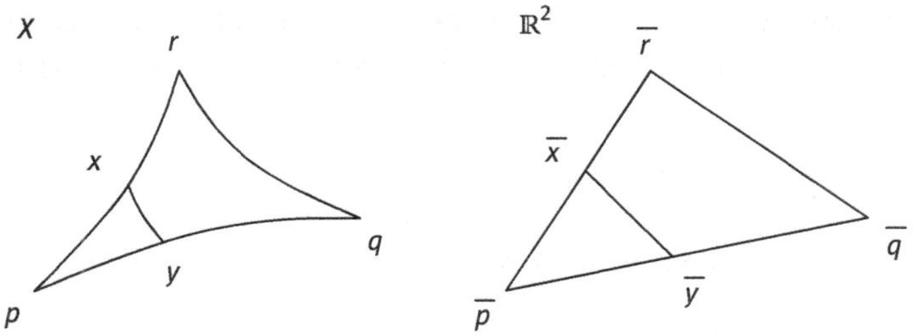

Fig. 1.2. Geodesic triangle (left) and its comparison triangle (right).

where $x_t := (1-t)q + tr$ for every $t \in [0,1]$. Inequality (1.2.2) says that the distance function in CAT(0) spaces is at least as convex as in the Euclidean plane. It is often applied with $t := \frac{1}{2}$ in which case $m := x_{\frac{1}{2}}$ is the midpoint of $[q,r]$ and we have

$$d(p,m)^2 \le \tfrac{1}{2}d(p,q)^2 + \tfrac{1}{2}d(p,r)^2 - \tfrac{1}{4}d(q,r)^2 . \tag{1.2.3}$$

As we shall see later, this inequality in turn implies CAT(0).

Let us record here that by Proposition 1.1.5 we have in a CAT(0) space (X,d) the following convexity property of the metric

$$d(x_t, y_t) \le (1-t)d(x_0, y_0) + td(x_1, y_1) \tag{1.2.4}$$

where $x, y\colon [0,1] \to X$ are geodesics. This is sometimes referred to as *joint convexity* of the metric. It in particular applies in the case $x_0 = x_1$.

Lemma 1.2.2. *Let (X,d) be a CAT(0) space and $x\colon [0,1] \to X$ be a geodesic. Then for every $t \in [0,1]$ the point $x_t := (1-t)x_0 + tx_1$ depends continuously on the endpoints x_0 and x_1.*

Proof. Choose one more geodesic $y\colon [0,1] \to X$. Applying (1.2.2) twice yields

$$\begin{aligned} d(x_t, y_t)^2 &\le (1-t)d(x_0, y_t)^2 + td(x_1, y_t)^2 - t(1-t)d(x_0, x_1)^2 \\ &\le (1-t)^2 d(x_0, y_0)^2 + t^2 d(x_1, y_1)^2 \\ &\quad + t(1-t)\left[d(x_0, y_1)^2 + d(x_1, y_0)^2 - d(x_0, x_1)^2 - d(y_0, y_1)^2\right] . \end{aligned}$$

The right hand side converges to 0 if $y_0 \to x_0$ and $y_1 \to x_1$. □

As a consequence of the preceding lemma we obtain that CAT(0) spaces are simply connected.

Definition 1.2.3 (Hadamard space). A complete CAT(0) space is called a *Hadamard space.*

The CAT(0) geometry is given by comparing geodesic triangles with the Euclidean ones. As a consequence, it is possible to compare geodesic quadrilaterals with quadrilaterals in the Euclidean space, which will be illuminated in Corollary 1.2.5 and Example 1.9. We now present a rather subtle inequality containing the quadruple comparison as a special case.

Proposition 1.2.4. *Let (X,d) be a CAT(0) space and*

$$x_0, x_1, y_0, y_1 \in X.$$

Then for every $t \in [0,1]$ and every parameter $\beta \in [0,1]$ we have

$$\begin{aligned} d(x_0, y_t)^2 + d(x_1, y_{1-t})^2 \le d(x_0, y_0)^2 + d(x_1, y_1)^2 + (2t^2 - t) d(y_0, y_1)^2 \\ + t d(x_0, x_1)^2 - t\beta \left[d(x_0, x_1) - d(y_0, y_1) \right]^2 \\ - t(1-\beta) \left[d(x_0, y_0) - d(x_1, y_1) \right]^2 , \end{aligned}$$

where, as always, $x_t := (1-t)x_0 + tx_1$ and $y_t := (1-t)y_0 + ty_1$.

Proof. Step 1: We first assume that $t = 1$ and $\beta = 0$. Denote $\gamma\colon [0,1] \to \mathcal{H}$ the geodesic connecting x_0 and y_1. Then (1.2.2) yields

$$\begin{aligned} d(y_0, \gamma_s)^2 \le (1-s) d(y_0, x_0)^2 + s d(y_0, y_1)^2 - s(1-s) d(x_0, y_1)^2 , \\ d(x_1, \gamma_s)^2 \le (1-s) d(x_1, x_0)^2 + s d(x_1, y_1)^2 - s(1-s) d(x_0, y_1)^2 , \end{aligned}$$

and furthermore, for every $\varepsilon > 0$,

$$\begin{aligned} d(x_1, y_0)^2 &\le \left[d(x_1, \gamma_s) + d(\gamma_s, y_0) \right]^2 \\ &\le (1+\varepsilon) d(x_1, \gamma_s)^2 + \left(1 + \tfrac{1}{\varepsilon}\right) d(\gamma_s, y_0)^2 \\ &\le (1+\varepsilon)(1-s) d(x_1, x_0)^2 + (1+\varepsilon) s d(x_1, y_1)^2 \\ &\quad + \left(1 + \tfrac{1}{\varepsilon}\right)(1-s) d(y_0, x_0)^2 + \left(1 + \tfrac{1}{\varepsilon}\right) s d(y_0, y_1)^2 \\ &\quad - \left(2 + \varepsilon + \tfrac{1}{\varepsilon}\right) s(1-s) d(x_0, y_1)^2 . \end{aligned}$$

Choosing $\varepsilon := \frac{s}{1-s}$ gives

$$\begin{aligned} d(x_0, y_1)^2 + d(x_1, y_0)^2 \le d(x_0, x_1)^2 + d(y_0, y_1)^2 \\ + \frac{1-s}{s} d(x_0, y_0)^2 + \frac{s}{1-s} d(x_1, y_1)^2 . \end{aligned}$$

If we further set $s := \frac{d(x_0, y_0)}{d(x_0, y_0) + d(x_1, y_1)}$, the last inequality becomes

$$\begin{aligned} d(x_0, y_1)^2 + d(x_1, y_0)^2 \le d(x_0, x_1)^2 + d(y_0, y_1)^2 + d(x_0, y_0)^2 + d(x_1, y_1)^2 \\ - \left[d(x_0, y_0) - d(x_1, y_1) \right]^2 , \end{aligned}$$

which shows the case $t = 1$ and $\beta = 0$.

Step 2: By a symmetry argument we also get

$$\begin{aligned} d(x_0, y_1)^2 + d(x_1, y_0)^2 \le d(x_0, x_1)^2 + d(y_0, y_1)^2 + d(x_0, y_0)^2 + d(x_1, y_1)^2 \\ - \left[d(x_0, x_1) - d(y_0, y_1) \right]^2 , \end{aligned}$$

and taking convex combinations yields the case of $t = 1$ and $\beta \in [0,1]$, namely,

$$d(x_0,y_1)^2 + d(x_1,y_0)^2 \le d(x_0,x_1)^2 + d(y_0,y_1)^2 + d(x_0,y_0)^2 + d(x_1,y_1)^2 - (1-\beta)\left[d(x_0,y_0) - d(x_1,y_1)\right]^2 - \beta\left[d(x_0,x_1) - d(y_0,y_1)\right]^2 . \tag{1.2.5}$$

Step 3: Finally, we obtain the general case applying (1.2.2) and (1.2.5) as follows

$$\begin{aligned} d(x_0,y_t)^2 + d(x_1,y_{1-t})^2 &\le (1-t)d(x_0,y_0)^2 + td(x_0,y_1)^2 \\ &\quad + (1-t)d(x_1,y_1)^2 + td(x_1,y_0)^2 \\ &\quad - 2t(1-t)d(y_0,y_1)^2 \\ &\le d(x_0,y_0)^2 + d(x_1,y_1)^2 + \left(2t^2 - t\right)d(y_0,y_1)^2 \\ &\quad + td(x_0,x_1)^2 - t\beta\left[d(x_0,x_1) - d(y_0,y_1)\right]^2 \\ &\quad - t(1-\beta)\left[d(x_0,y_0) - d(x_1,y_1)\right]^2 . \end{aligned}$$

The proof is now complete. □

Corollary 1.2.5. *Let (X,d) be a CAT(0) space. Then for every four points $x_0, x_1, y_0, y_1 \in X$ we have*

$$d(x_0,y_1)^2 + d(x_1,y_0)^2 \le d(x_0,y_0)^2 + d(x_1,y_1)^2 + 2d(x_0,x_1)\,d(y_0,y_1),$$

and in particular,

$$d(x_0,y_1)^2 + d(x_1,y_0)^2 \le d(x_0,y_0)^2 + d(x_1,y_1)^2 + d(x_0,x_1)^2 + d(y_0,y_1)^2 .$$

Proof. Follows directly from Proposition 1.2.4. If (X,d) is complete, we can alternatively use Theorem 2.3.5 (iv) with

$$s := t := \frac{d(x_3,x_4)}{d(x_1,x_2) + d(x_3,x_4)}$$

to obtain the desired inequalities. □

The second inequality in Corollary 1.2.5 can be actually strengthened; see Exercise 1.9, where we also present its geometrical meaning.

Angles in geodesic spaces

The notion of an angle in Hadamard spaces is more important in geometry than analysis. We will see in Theorem 1.3.3 that the CAT(0) property can be expressed in terms of angles.

Let first a, b, c be three points in the Euclidean plane $\mathbb{R}^2$. The angle between the line segments $[a,b]$ and $[a,c]$ will be denoted $\measuredangle(b,a,c)$, that is,

$$\measuredangle(b,a,c) := \arccos \frac{\langle b-a, c-a\rangle}{\|b-a\|\|c-a\|} .$$

Definition 1.2.6 (Angle between two geodesics). Let now (X, d) be a geodesic space. We define the *angle* between two geodesics $\gamma\colon [0,1] \to X$ and $\eta\colon [0,1] \to X$ with $\gamma_0 = \eta_0$ by

$$\alpha(\gamma, \eta) := \limsup_{s,t \to 0+} \measuredangle\left(\overline{\gamma_s}, \overline{\gamma_0}, \overline{\eta_t}\right),$$

where $\overline{\gamma_s}, \overline{\gamma_0}, \overline{\eta_t}$ are the vertices of the comparison triangle $\triangle(\overline{\gamma_s}, \overline{\gamma_0}, \overline{\eta_t})$ for the geodesic triangle $\triangle(\gamma_s, \gamma_0, \eta_t)$. Note that $\overline{\gamma_s}$ is *not* a comparison point for γ_s in the comparison triangle $\triangle(\overline{\gamma_0}, \overline{\gamma_1}, \overline{\eta_1})$, that is, $\overline{\gamma_s}$ is not $(1-s)\overline{\gamma_0} + s\overline{\gamma_1}$. And likewise for $\overline{\eta_t}$.

An angle is hence a number from $[0, \pi]$. Let now $(\mathcal{H}, d)$ be a Hadamard space and $\gamma\colon [0,1] \to \mathcal{H}$ and $\eta\colon [0,1] \to \mathcal{H}$ be geodesics with $\gamma_0 = \eta_0$. Without loss of generality we may assume $d(\gamma_0, \gamma_1) = d(\eta_0, \eta_1)$. Since both functions $s \mapsto \measuredangle(\overline{\gamma_s}, \overline{\gamma_0}, \overline{\eta_t})$ and $t \mapsto \measuredangle(\overline{\gamma_s}, \overline{\gamma_0}, \overline{\eta_t})$ are then nondecreasing, we have

$$\alpha(\gamma, \eta) = \lim_{t \to 0+} \measuredangle\left(\overline{\gamma_t}, \overline{\gamma_0}, \overline{\eta_t}\right)$$

and by an easy trigonometry,

$$= \lim_{t \to 0+} 2 \arcsin \frac{d(\gamma_t, \eta_t)}{2t d(\gamma_0, \gamma_1)}.$$

Given points $p, q, r \in \mathcal{H}$, we will denote the angle between the geodesics $[p, q]$ and $[p, r]$ by $\alpha(q, p, r)$, and the corresponding angle in the comparison triangle by $\overline{\alpha}(q, p, r)$, that is, $\overline{\alpha}(q, p, r) = \measuredangle(\overline{q}, \overline{p}, \overline{r})$, where $\overline{p}, \overline{q}, \overline{r}$ are the vertices of a comparison triangle for the geodesic triangle $\triangle(p, q, r)$. Angles also behave nicely when we perturb the geodesics. To show this, we need the following lemma.

Lemma 1.2.7. *Let (X, d) be a geodesic space and $\gamma_1, \gamma_2, \gamma_3$ be three geodesics issuing from the same point. Then*

$$\alpha(\gamma_1, \gamma_2) \le \alpha(\gamma_1, \gamma_3) + \alpha(\gamma_3, \gamma_2).$$

Proof. Exercise 1.10. □

Now we can prove the continuity of an angle.

Proposition 1.2.8. *Let $(\mathcal{H}, d)$ be a Hadamard space. Then:*
(i) *The function $\alpha(\cdot, \cdot, \cdot)$ is upper semicontinuous on $\mathcal{H}^3$.*
(ii) *For a fixed $p \in \mathcal{H}$ the function $\alpha(\cdot, p, \cdot)$ is continuous on $\mathcal{H}^2$.*

Proof. (i) Choose sequences (x_n), (y_n) and (p_n) in $\mathcal{H}$ converging to points x, y and p, respectively. For $t \in (0, 1]$ put $a(t) := \overline{\alpha}((1-t)p + tx, p, (1-t)p + ty)$ and $a_n(t) := \overline{\alpha}((1-t)p_n + tx_n, p_n, (1-t)p_n + ty_n)$. We already know that $a(t)$ is a nondecreasing function of t and $\alpha(x, p, y) = \lim_{t \to 0+} a(t)$ along with $\alpha(x_n, p_n, y_n) = \lim_{t \to 0+} a_n(t)$ for each $n \in \mathbb{N}$. Obviously, for a fixed $t \in (0, 1]$ we have $a_n(t) \to a(t)$ as $n \to \infty$. Given $\varepsilon > 0$, there exists $\tau \in (0, 1]$ such that $a(t) - \varepsilon \le \alpha(x, p, y)$ for every $t \in (0, \tau]$. Then for n big enough, we have $a_n(\tau) \le a(\tau) + \varepsilon$, and therefore

$$\alpha(x_n, p_n, y_n) \le a_n(\tau) \le a(\tau) + \varepsilon \le \alpha(x, p, y) + 2\varepsilon.$$

Thus $\limsup \alpha(x_n, p_n, y_n) \le \alpha(x, p, y)$.

(ii) We keep the notation, but assume $p_n = p$ for each $n \in \mathbb{N}$. Applying Lemma 1.2.7 yields

$$\left|\alpha\left(x_n, p, y_n\right) - \alpha\left(x, p, y\right)\right| \leq \alpha\left(x, p, x_n\right) + \alpha\left(y, p, y_n\right),$$

which, along with the facts that $\lim_n \alpha(x, p, x_n) = 0$ and $\lim_n \alpha(y, p, y_n) = 0$, gives $\lim_n \alpha(x_n, p, y_n) = \alpha(x, p, y)$. □

Examples of Hadamard spaces

Let us now list classical examples of Hadamard spaces. For detailed constructions and proofs, the reader is referred to the Bridson–Haefliger book [51]. We start with two basic examples of Hadamard spaces, which in some sense represent the most extreme cases: curvature 0 and curvature $-\infty$.

Example 1.2.9 (Hilbert spaces). Hilbert spaces are Hadamard. The geodesics are the line segments. Moreover, it is known that a Banach space is CAT(0) if and only if it is Hilbert; see Exercise 1.4.

Example 1.2.10 ($\mathbb{R}$-trees). Recall that a metric space (X, d) is an $\mathbb{R}$-tree if it is uniquely geodesic and for every $x, y, z \in X$ we have $[x, z] = [x, y] \cup [y, z]$ whenever $[x, y] \cap [y, z] = \{y\}$. Notice that all triangles in an $\mathbb{R}$-tree are trivial.

Hadamard manifolds

The most illuminating instances of Hadamard spaces are Hadamard manifolds. By definition, a Hadamard manifold is a complete simply connected Riemannian manifold of nonpositive sectional curvature. The class of Hadamard manifolds includes:

Example 1.2.11 (Hyperbolic spaces). We equip $\mathbb{R}^{n+1}$ with the $(-1, n)$-inner product

$$\langle x, y\rangle_{(-1,n)} := -x^0 y^0 + \sum_{i=1}^{n} x^i y^i$$

for $x := (x^0, x^1, \dots, x^n)$ and $y := (y^0, y^1, \dots, y^n)$. Define

$$\mathbb{H}^n := \left\{x \in \mathbb{R}^{n+1} : \langle x, x\rangle_{(-1,n)} = -1, x_0 > 0\right\}.$$

Then $\langle\cdot\,,\cdot\rangle$ induces a Riemannian metric g on the tangent spaces $T_p\mathbb{H}^n \subset T_p\mathbb{R}^{n+1}$ for $p \in \mathbb{H}^n$. The sectional curvature of $(\mathbb{H}^n, g)$ is -1 at every point.

Example 1.2.12 (Manifolds of positive definite matrices). The space $P(n, \mathbb{R})$ of symmetric positive definite matrices $n \times n$ with real entries is a Hadamard manifold if it is equipped with the Riemannian metric

$$\langle X, Y\rangle_A := \operatorname{Tr}\left(A^{-1}XA^{-1}Y\right), \quad X, Y \in T_A\left(P(n, \mathbb{R})\right),$$

for every $A \in P(n, \mathbb{R})$.

The class of Hadamard manifolds also contains symmetric spaces of noncompact type, like $SL(n,\mathbb{R})/SO(n,\mathbb{R})$. Note that $SL(n,\mathbb{R})/SO(n,\mathbb{R})$ can be identified with the submanifold $P_1(n,\mathbb{R}) := \{A \in P(n,\mathbb{R})\colon \det A = 1\}$.

As far as *infinite* dimensional manifolds (that is, those modeled on a Hilbert space) are concerned, the Hopf–Rinow theorem is not valid any more and we do not have the existence of geodesics in general. However, if the manifold in question is Hadamard, we obtain the existence of geodesics as a consequence of nonpositive curvature. Thus infinite dimensional Hadamard manifolds are another instance of Hadamard spaces; see [125, XI. Proposition 3.4.]. The complex Hilbert ball with the hyperbolic metric is a prominent example of an infinite dimensional Hadamard manifold and we will describe it in detail:

Example 1.2.13 (Hilbert ball). Let $(H, \|\cdot\|)$ be a complex Hilbert space with inner product $\langle\cdot,\cdot\rangle$ and put $\mathbb{B} := \{x \in H\colon \|x\| < 1\}$. We equip $\mathbb{B}$ with the hyperbolic metric

$$\rho(x,y) := \tanh^{-1}\sqrt{1-\sigma(x,y)}\,, \qquad x,y \in \mathbb{B}\,,$$

where

$$\sigma(x,y) := \frac{\left(1-\|x\|^2\right)\left(1-\|y\|^2\right)}{1-\langle x,y\rangle}\,, \qquad x,y \in \mathbb{B}\,.$$

The metric space $(\mathbb{B},\rho)$ is called *the Hilbert ball*. For each $a \in \mathbb{B}$, the *Möbius transformation* associated with a is defined by

$$M_a(x) := \left(\sqrt{1-\|a\|^2}Q_a + P_a\right)\frac{x+a}{1-\langle x,a\rangle}\,, \qquad x \in \mathbb{B}\,,$$

where P_a is the projection onto the one-dimensional space span(a) and $Q_a := I - P_a$, with I being the identity mapping. Möbius transformations are automorphisms on $\mathbb{B}$, and we have $M_{-a}(a) = 0$ for each $a \in \mathbb{B}$. This will enable us in the sequel, given a point $x \in \mathbb{B}$, to consider $x = 0$. The Hilbert ball is a uniquely geodesic space and each pair of points lies on a geodesic *line*. It also has a structure of an infinite dimensional Riemannian manifolds. Indeed, to each a point $x \in \mathbb{B}$, we associate a Hilbert space H_x, whose elements will be denoted by (x,y), where $y \in \mathbb{B}$. The space H_x is equipped with the inner product

$$\langle (x,y),(x,z)\rangle_{H_x} := \frac{\rho(x,y)\rho(x,z)}{\|M_{-x}(y)\|\,\|M_{-x}(z)\|}\langle M_{-x}(y), M_{-x}(z)\rangle\,,$$

for $x \neq y$ and $x \neq z$ from $\mathbb{B}$. The symbol $\|\cdot\|_{H_x}$ will then stand for the corresponding norm. For more details, see [92].

We will show that $(\mathbb{B},\rho)$ is a Hadamard space in a few steps. *Step 1:* We claim first it is a Busemann space, that is,

$$\rho\left(m_1,m_2\right) \leq \tfrac{1}{2}\rho(x,y)\,, \tag{1.2.6}$$

for every $x, y, z \in \mathbb{B}$, with m_1, m_2 being the midpoints of $[x, y]$ and $[x, z]$, respectively. Without loss of generality (by virtue of a Möbius transformation), we may assume $z = 0$. Then

$$x = \frac{2m_1}{1 + \|m_1\|^2}, \quad y = \frac{2m_2}{1 + \|m_2\|^2},$$

and one can readily check that the desired inequality (1.2.6) is equivalent to

$$\sigma(x, y) \leq \frac{\sigma(m_1, m_2)^2}{[2 - \sigma(m_1, m_2)]^2}. \tag{1.2.7}$$

To see that the last inequality indeed holds true, let us compute

$$\begin{aligned}\sigma(x, y) &= \frac{\left(1 - \|m_1\|^2\right)^2 \left(1 - \|m_2\|^2\right)^2}{\left|\left(1 - \|m_1\|^2\right)\left(1 - \|m_2\|^2\right) + 2\left(\|m_1\|^2 + \|m_2\|^2 - 2\langle m_1, m_2\rangle\right)\right|^2} \\ &\leq \frac{\left(1 - \|m_1\|^2\right)^2 \left(1 - \|m_2\|^2\right)^2}{\left[\left(1 - \|m_1\|^2\right)\left(1 - \|m_2\|^2\right) + 2\|m_1 - m_2\|^2\right]^2} \\ &\leq \frac{\sigma(m_1, m_2)^2}{\left[\sigma(m_1, m_2) + \frac{2\|m_1 - m_2\|^2}{|1 - \langle m_1, m_2\rangle|^2}\right]^2} \\ &\leq \frac{\sigma(m_1, m_2)^2}{[2 - \sigma(m_1, m_2)]^2},\end{aligned}$$

which is exactly (1.2.7). The Hilbert ball is thus a Busemann space. *Step 2:* Next we show that

$$\rho(y, z)^2 \geq \rho(x, y)^2 + \rho(x, z)^2 - 2\operatorname{Re}\langle (x, y), (x, z)\rangle \tag{1.2.8}$$

holds for every $x, y, z \in \mathbb{B}$. To this end, put $y_t := (1 - t)x + ty$ and $z_t := (1 - t)x + tz$, for each $t \in [0, 1]$. By the previous step, we know the Hilbert ball is a Busemann space and we can employ Proposition 1.1.5 to conclude that $\rho(y_t, z_t)^2 \leq t^2 \rho(y, z)^2$ for each $t \in [0, 1]$ and hence it suffices to show

$$\lim_{t \to 0} \frac{\rho(y_t, z_t)^2}{t^2} \geq \rho(x, y)^2 + \rho(x, z)^2 - 2\operatorname{Re}\langle (x, y), (x, z)\rangle.$$

Without loss of generality, we may assume $y \neq x$, $z \neq x$ and $x = 0$. We then have

$$\rho(y_t, z_t) = \rho\left(\tanh\left(t \tanh^{-1}\|y\|\right)\frac{y}{\|y\|}, \tanh\left(t \tanh^{-1}\|z\|\right)\frac{z}{\|z\|}\right).$$

Since ρ and the Hilbert space metric are locally isometric, one can write

$$\begin{aligned}\lim_{t\to 0}\frac{\rho(y_t,z_t)^2}{t^2} &= \lim_{t\to 0}\frac{1}{t^2}\left\|\tanh\left(t\tanh^{-1}\|y\|\right)\frac{y}{\|y\|}-\tanh\left(t\tanh^{-1}\|z\|\right)\frac{z}{\|z\|}\right\| \\ &= \left\|\tanh^{-1}(\|y\|)\frac{y}{\|y\|}-\tanh^{-1}(\|z\|)\frac{z}{\|z\|}\right\| \\ &= \|(0,y)-(0,z)\|_{H_0}^2 \\ &= \rho(x,y)^2+\rho(x,z)^2-2\operatorname{Re}\langle(x,y),(x,z)\rangle ,\end{aligned}$$

which proves (1.2.8). *Step 3:* Finally, we will show that the Hilbert ball satisfies (1.2.2), that is,

$$\rho(y,x_t)^2 \le (1-t)\rho(y,x_0)^2 + t\rho(y,x_1)^2 - t(1-t)\rho(x_0,x_1)^2 , \tag{1.2.9}$$

for each geodesic $x\colon [0,1] \to \mathbb{B}$ and $y \in \mathbb{B}$. Applying (1.2.8) with $x := x_t,\ y := x_0,$ and $z := y$, we get

$$\rho(x_0,y)^2 \ge \rho(x_t,x_0)^2 + \rho(x_t,y)^2 - 2\operatorname{Re}\langle(x_t,x_0),(x_t,y)\rangle ,$$

and applying (1.2.8) again with $x := x_t,\ y := x_1$, and $z := y$, we get

$$\rho(x_1,y)^2 \ge \rho(x_t,x_1)^2 + \rho(x_t,y)^2 - 2\operatorname{Re}\langle(x_t,x_1),(x_t,y)\rangle .$$

Now multiply the last inequality by t and the second last inequality by $(1-t)$ and sum them up to obtain (1.2.9). Thus the Hilbert ball is a CAT(0) space. Since $(\mathbb{B},\rho)$ is obviously complete, we can conclude it is a Hadamard space.

A more subtle construction of an infinite-dimensional *weak* Riemannian manifold, whose metric completion with respect to the L^2-metric is a Hadamard space, is due to B. Clarke [63]; see also [64, 65]. This manifold arises as the space of Riemannian metrics on a given closed (finite dimensional) manifold. Note however that the completion is no longer a manifold. This construction has a complex counterpart in Kähler geometry, where the metric completion with respect to the L^2-metric of the space of Kähler potentials in a given cohomology class is also a Hadamard space [191], which will be revisited in Example 5.1.17. An analogous (and earlier) result due to S. Yamada [208] states that the completion of the Teichmüller space with respect to the Weil–Petersson metric is a Hadamard space; see also [207] for a nice survey.

Complexes

A significant part of CAT(0) geometry and geometric group theory is devoted to CAT(0) complexes, simplicial and cubical complexes being the most important instances. One of the prominent classes of CAT(0) complexes is called *Euclidean buildings*, which have been thoroughly treated in [1].

In Chapter 8 we shall present the BHV tree space, which is a nice example of a CAT(0) cubical complex. The following two facts are recorded here for future reference.

Theorem 1.2.14 (Bridson's theorem). *If a cubical complex is finite dimensional or locally finite, then it is a complete geodesic space.*

There exists a useful criterion whether a cubical complex is CAT(0) due to M. Gromov.

Theorem 1.2.15. *A finite dimensional cubical complex is a Hadamard space if and only if it is simply connected and the link of each of its vertices is a flag complex.*

As an authoritative text on the subject, we recommend [51] by M. Bridson and A. Haefliger.

Constructions of Hadamard spaces

After presenting examples of Hadamard spaces, we now turn to various techniques which enable us to construct new Hadamard spaces out of known ones. We start with the definition of convexity, which will be further explored in Chapter 2. Let $(\mathcal{H}, d)$ be a Hadamard space. A set $C \subset \mathcal{H}$ is *convex* if, given $x, y \in C$, we have $[x, y] \subset C$. We say that a function $f\colon \mathcal{H} \to (-\infty, \infty]$ is *convex* if, for every geodesic $\gamma\colon [0, 1] \to \mathcal{H}$, the function $f \circ \gamma$ is convex, that is,

$$f(\gamma_t) \leq (1-t) f(\gamma_0) + t f(\gamma_1),$$

for every $t \in (0, 1)$.

Subspaces

A closed convex subset of a Hadamard space with the restricted metric is itself a Hadamard space.

Gluing along convex sets

Let $(\mathcal{H}_i, d_i)_{i \in I}$ be a family of Hadamard spaces (with an arbitrary index set I) and C be a Hadamard space. Suppose $C_i \subset \mathcal{H}_i$ are convex closed and $F_i\colon C \to C_i$ are isometries, for each $i \in I$. Define the equivalence relation $F_i(x) \sim F_j(x)$ for every $x \in C$ and $i, j \in I$. Then the quotient of the disjoint union

$$\mathcal{H} := \bigsqcup_{i \in I} \mathcal{H}_i / \sim$$

is said to be constructed *by gluing* of $(\mathcal{H}_i, d_i)$ along C and it is a Hadamard space.

Products

Let $(\mathcal{H}_i, d_i)$ be a Hadamard space for every $i = 1, \ldots, N$, where $N \in \mathbb{N}$. Define $\mathcal{H} := \mathcal{H}_1 \times \cdots \times \mathcal{H}_N$ and for every $x := (x_1, \ldots, x_N)$, $y := (y_1, \ldots, y_N) \in \mathcal{H}$ put

$$d(x, y) := \left(\sum_{i=1}^{N} d_i (x_i, y_i)^2 \right)^{\frac{1}{2}}.$$

Then $(\mathcal{H}, d)$ is a Hadamard space.

We can extend this construction for an infinite index set and add weights. Assume $(\mathcal{H}_i, d_i, o_i)$ are pointed metric spaces and $w_i > 0$ for each $i \in I$, where I is an arbitrary index set. Then the *weighted product* of these spaces is defined as

$$\mathcal{H} := \left\{ x \in \prod_{i \in I} \mathcal{H}_i : \sum_{i \in I} w_i d_i (x_i, o_i)^2 < \infty \right\}, \tag{1.2.10}$$

with the metric

$$d(x, y) := \left(\sum_{i \in I} w_i d_i (x_i, y_i)^2 \right)^{\frac{1}{2}}.$$

It is easy to show that $(\mathcal{H}, d)$ is a Hadamard space if all the factors are such; see Exercise 1.8.

Warped products

Let $(\mathcal{H}_1, d_1)$ and $(\mathcal{H}_2, d_2)$ be Hadamard spaces and $f : \mathcal{H}_1 \to (0, \infty)$ be convex continuous function. The length of a path (β_1, β_2) in $\mathcal{H}_1 \times \mathcal{H}_2$ is given as

$$\sup \sum_{i=1}^{n} \left[d_1 (\beta_1 (t_{i-1}), \beta_1 (t_i))^2 + f(\beta_1 (t_{i-1}))^2 d_2 (\beta_2 (t_{i-1}), \beta_2 (t_i))^2 \right]^{\frac{1}{2}},$$

where we take supremum over the set of all partitions $0 = t_0 < \cdots < t_n = 1$ of $[0, 1]$, with an arbitrary $n \in \mathbb{N}$. We then define the distance between two points in $\mathcal{H}_1 \times \mathcal{H}_2$ as the infimum of path-lengths between these two points. The resulting space, called the *warped product* of $\mathcal{H}_1$ and $\mathcal{H}_2$ with respect to the function f, is a Hadamard space and is denoted $\mathcal{H}_1 \times_f \mathcal{H}_2$.

Ultraproducts and ultrapowers

A finitely additive probability measure ω on $\mathbb{N}$ is called a *nonprincipal ultrafilter* on $\mathbb{N}$ if every set $S \subset \mathbb{N}$ is measurable, the value of $\omega(S)$ is either 0 or 1, and $\omega(S) = 0$ whenever S is finite. The existence of nonprincipal ultrafilters is a standard result, which follows from Zorn's lemma[2]. There is however no distinguished nonprincipal

2 We assume Zorn's lemma here. In this connection recall Jerry Bona's famous quote: "The axiom of choice is obviously true, the well-ordering principle obviously false, and who can tell about Zorn's lemma?"

ultrafilter on $\mathbb{N}$ and we can therefore choose an arbitrary one, let us denote it ω. On the other hand one must remember that all the objects in the following construction do depend on this choice.

Ultralimits are limits with respect to a nonprincipal ultrafilter, more precisely, given a bounded sequence $(a_n) \subset \mathbb{R}$, there exists a unique number $l \in \mathbb{R}$ such that

$$\omega \{ n \in \mathbb{N} \colon |a_n - l| < \varepsilon \} = 1 ,$$

for every $\varepsilon > 0$. We say l is the *ultralimit* of (a_n) and use the notation $l = \omega\text{-lim}_n a_n$.

Let $(X_n, d_n, o_n)_{n \in \mathbb{N}}$ be a sequence of pointed metric spaces. Let X denote the set of sequences (x_n), where $x_n \in X_n$ and $\sup_{n \in \mathbb{N}} d_n(x_n, o_n) < \infty$. Next define

$$x \sim y \quad \text{if } \omega\text{-}\lim_n d_n (x_n, y_n) = 0,$$

which is an equivalence relation on X. Denote the quotient $X_\omega = X/\sim$ and put

$$d_\omega ((x_n), (y_n)) := \omega\text{-}\lim_n d_n (x_n, y_n) .$$

Then (X_ω, d_ω) is a complete metric space which we call the *ultralimit*, or *ultraproduct*, of (X_n, d_n, o_n). For a constant sequence of spaces, we use the term *ultrapower*. If all (X_n, d_n) are length spaces (geodesic spaces, respectively), so is the ultralimit. The ultralimit of a sequence of pointed CAT(0) spaces is a Hadamard space.

Example 1.2.16 (Asymptotic cone). Let (X, d) be a metric space and $o \in X$. The ultralimit of the sequence $(X, \frac{1}{n} d, o)$ is called the *asymptotic cone* of X at the point o.

Gromov–Hausdorff limits

Recall that the *Gromov–Hausdorff* distance between compact metric spaces X, Y is

$$d_{\mathrm{GH}}(X, Y) := \inf d_{\mathrm{H}} (f(X), g(Y)) ,$$

where we take inf over all metric spaces Z and all isometries $f \colon X \to Z$ and $g \colon Y \to Z$. The convergence induced by d_{GH} is called *Gromov–Hausdorff*. Note X and Y are isometric if and only if $d_{\mathrm{GH}}(X, Y) = 0$. The *pointed Gromov–Hausdorff convergence* is an appropriate analog of the Gromov–Hausdorff convergence for *locally* compact spaces. A sequence of pointed metric spaces (X_n, d_n, o_n) converges to (X, d, o) if for every $r > 0$ we have

$$d_{\mathrm{GH}} (B(o_n, r), B(o, r)) \to 0 .$$

Both the Gromov–Hausdorff and pointed Gromov–Hausdorff convergences preserve the property of being Hadamard. In particular, this applies to a sequence of Hadamard manifolds, which can be actually regarded as one of the motivations for defining Hadamard spaces.

If (X_n, d_n, o_n) is a sequence of locally compact metric spaces, then its pointed Gromov–Hausdorff limit is isometric to its ultralimit with respect to an arbitrary ultrafilter. Ultralimits are hence a faithful generalization of the pointed Gromov–Hausdorff convergence for spaces which are not locally compact. For another such generalization called an *asymptotic relation*, see [124, Definition 3.1].

Universal coverings

We will now show how to construct a Hadamard space out of a locally Hadamard space. Recall that a metric space is *locally Hadamard* if it is connected and each of its points has a neighborhood which is a Hadamard space.

Let X be a connected topological space. A connected and simply connected topological space $\tilde{X}$ is called a *universal covering* of X if there exists a continuous surjective mapping $\pi\colon \tilde{X} \to X$ such that each $x \in X$ has an open neighborhood U with the property that $\pi^{-1}(U)$ is a disjoint union of open sets in $\tilde{X}$ each of which is mapped homeomorphically onto U by π. A universal covering is unique. Given a path $\gamma\colon [0,1] \to X$ and point $x_0 \in \pi^{-1}(\gamma_0)$, there exists a unique path $\tilde{\gamma}\colon [0,1] \to \tilde{X}$ such that $\tilde{\gamma}_0 = x_0$ and $\pi(\tilde{\gamma}_t) = \gamma_t$ for $t \in [0,1]$. The path $\tilde{\gamma}$ is called a *lift* of γ through x_0. We define the distance $\tilde{d}$ to be the induced length metric on $\tilde{X}$.

Hadamard spaces also arise as universal coverings of locally Hadamard spaces by a version of the Cartan–Hadamard theorem [4]. More precisely, if $(\mathcal{H}, d)$ is a locally Hadamard space, then its universal covering $(\tilde{\mathcal{H}}, \tilde{d})$ is a Hadamard space.

Tangent cones

Unlike Riemannian manifolds, a geodesic metric space has no tangent spaces. It is however possible to construct a *tangent cone* at a given point. It turns out that a (completed) tangent cone of a locally Hadamard space is again a Hadamard space. Let us start with the notion of a Euclidean cone over a metric space.

Given a metric space (X, d), we define the *Euclidean cone* over X by

$$\operatorname{cone} X := ([0,\infty) \times X) / \sim,$$

where the equivalence relation $\sim$ is defined by $(s,x) \sim (t,y)$ if $(s = t = 0)$ or $(s = t$ and $x = y)$. For a pair of points (s,x) and (t,y) in $\operatorname{cone} X$, we set

$$d_{\text{cone}}((s,x),(t,y)) := \left[s^2 + t^2 - 2st\cos\left(\min\left\{d(x,y),\pi\right\}\right)\right]^{\frac{1}{2}}. \tag{1.2.11}$$

Then $(\operatorname{cone} X, d_{\text{cone}})$ is a metric space which is complete if and only (X, d) is such.

Next we define the space of directions. Let now (X, d) be a geodesic space and fix a point $p \in X$. Two geodesics γ, η issuing from p are said to determine the same direction if their angle $\alpha(\gamma, \eta) = 0$. By Lemma 1.2.7, the angle determines an equivalence relation and the *space of directions* at the point p is the quotient of the set of geodesics issuing from p with respect to this equivalence relation. Lemma 1.2.7 also gives that $\alpha(\cdot,\cdot)$ is a metric on the space of directions. We denote the space of directions at p by $\operatorname{dir}_p X$ and its completion with respect to the metric α by $\overline{\operatorname{dir}}_p X$.

Finally, we define the *tangent cone* at p as the Euclidean cone over $\operatorname{dir}_p X$, that is $T_p X := \operatorname{cone}(\operatorname{dir}_p X)$, and the *completed tangent cone* at p as the Euclidean cone over $\overline{\operatorname{dir}}_p X$, that is $\overline{T}_p X := \operatorname{cone}(\overline{\operatorname{dir}}_p X)$. As a matter of fact, the completed tangent cone at p can be equivalently obtained as the completion of the tangent cone T_p with respect to the cone metric (1.2.11). This metric on $T_p X$ or $\overline{T}_p X$ is usually denoted by d_p.

Theorem 1.2.17. *Let $(\mathcal{H}, d)$ be a locally Hadamard space. The completed tangent cone $\overline{T}_p\mathcal{H}$ at a point $p \in \mathcal{H}$ is a Hadamard space.*

The previous theorem will be used in the proof of Theorem 4.1.2. For further information on tangent cones and the proof of the above theorem, see [51].

Nonlinear Lebesgue spaces

We will now define $\mathcal{L}^p(\Omega, \mathcal{H})$ spaces of mapping from a probability space $(\Omega, \mathcal{F}, \mu)$ to a Hadamard space $(\mathcal{H}, d)$. Here $\mathcal{F}$ denotes a σ-algebra on Ω and μ is a probability measure. On $\mathcal{H}$ we consider the Borel σ-algebra. Given two measurable maps $f\colon \Omega \to \mathcal{H}$ and $g\colon \Omega \to \mathcal{H}$, we put

$$d_p(f,g) := \begin{cases} \left[\int_\Omega d\left(f(\omega), g(\omega)\right)^p \,d\mu(\omega)\right]^{\frac{1}{p}}, & \text{if } p \in [1,\infty), \\ \operatorname{ess\,sup}_{\omega\in\Omega} d\left(f(\omega), g(\omega)\right), & \text{if } p = \infty. \end{cases}$$

The space $\mathcal{H}$ can be identified with constant maps from Ω to $\mathcal{H}$. We will say that an $\mathcal{F}$-measurable map $f\colon \Omega \to \mathcal{H}$ belongs to the space $\mathcal{L}^p(\Omega, \mathcal{H})$ if

$$d_p(f,x) < \infty,$$

for some (or equivalently any) point $x \in \mathcal{H}$. More precisely, the space $\mathcal{L}^p(\Omega, \mathcal{H})$ contains *equivalence classes* of such maps, that is, we take the quotient with respect to

$$f \sim g \quad \text{if } d_p(f,g) = 0.$$

Then d_p becomes a metric on $\mathcal{L}^p(\Omega, \mathcal{H})$. Notice that the space $\mathcal{L}^p(\Omega, \mathcal{H})$ of course depends on Ω, $\mathcal{F}$, μ and $\mathcal{H}$, and d, but we explicitly write only Ω and $\mathcal{H}$ if no ambiguity is possible. In Section 7.3, we will need to specify the σ-algebra as well.

Proposition 1.2.18. *The metric space $(\mathcal{L}^p(\Omega, \mathcal{H}), d_p)$ is complete. If $p = 2$, it is Hadamard.*

Proof. Step 1: We first show completeness. Let (f_n) be a Cauchy sequence in $\mathcal{L}^p(\Omega, \mathcal{H})$. Select a subsequence such that

$$d_p\left(f_{n_k}, f_{n_{k+1}}\right) \le 2^{-k-1},$$

for every $k \in \mathbb{N}$, and define the mapping $g_l(\omega) := \sum_{k=l}^{\infty} d(f_{n_k}(\omega), f_{n_{k+1}}(\omega))$, for each $l \in \mathbb{N}$ and almost every $\omega \in \Omega$. Then $g_l \colon \Omega \to [0, \infty]$ and its L^p-norm satisfies

$$\|g_l\|_{L^p} \leq \sum_{k=l}^{\infty} d_p\left(f_{n_k}, f_{n_{k+1}}\right) \leq \sum_{k=l}^{\infty} 2^{-k-1} = 2^{-l}.$$

Thus $g_l \xrightarrow{L^p} 0$ and consequently $g_l \to 0$ almost everywhere on Ω. Since

$$d\left(f_{n_j}(\omega), f_{n_l}(\omega)\right) \leq \sum_{k=l}^{j-1} d\left(f_{n_k}(\omega), f_{n_{k+1}}(\omega)\right) \leq g_l(\omega),$$

for almost every $\omega \in \Omega$, the sequence $(f_{n_l}(\omega))$ is Cauchy. There exists a point $f(\omega) \in \mathcal{H}$ such that $f_{n_l}(\omega) \to f(\omega)$ as $l \to \infty$, for almost every $\omega \in \Omega$. Moreover, we have

$$d\left(f(\omega), f_{n_l}(\omega)\right) \leq g_l(\omega) \to 0,$$

which implies that $f \colon \Omega \to \mathcal{H}$ is measurable and also $d_p(f, f_{n_l}) \leq \|g_l\|_{L^p} \to 0$. Thus $f_{n_l} \to f$ in $\mathcal{L}^p(\Omega, \mathcal{H})$, and since (f_n) is Cauchy, also $f_n \to f$ in $\mathcal{L}^p(\Omega, \mathcal{H})$. It shows the first statement.

Step 2: Let us now consider the case $p = 2$. For $f_0, f_1 \in \mathcal{L}^2(\Omega, \mathcal{H})$ and almost every $\omega \in \Omega$, let $t \mapsto f_t(\omega)$ be the geodesic connecting $f_0(\omega)$ and $f_1(\omega)$, hence

$$d\left(f_s(\omega), f_t(\omega)\right) = |t - s| d\left(f_0(\omega), f_1(\omega)\right), \tag{1.2.12}$$

for every $s, t \in [0, 1]$. By Lemma 1.2.2, the geodesic depends continuously on the endpoints $f_0(\omega)$ and $f_1(\omega)$, and thus in a measurable way on ω. We can integrate (1.2.12) to get

$$d_2\left(f_s, f_t\right) = |t - s| d_2\left(f_0, f_1\right).$$

This implies that, for each $t \in [0, 1]$, the mapping $f_t \colon \Omega \to \mathcal{H}$ belongs to $\mathcal{L}^2(\Omega, \mathcal{H})$ and $t \mapsto f_t$ is a geodesic in $\mathcal{L}^2(\Omega, \mathcal{H})$. For $g \in \mathcal{L}^2(\Omega, \mathcal{H})$ and almost every $\omega \in \Omega$, we have by (1.2.2) that

$$\begin{aligned} d\left(g(\omega), f_t(\omega)\right)^2 \leq (1-t) d\left(g(\omega), f_0(\omega)\right)^2 + t d\left(g(\omega), f_1(\omega)\right)^2 \\ - t(1-t) d\left(f_0(\omega), f_1(\omega)\right)^2, \end{aligned}$$

and after integration,

$$d_2\left(g, f_t\right)^2 \leq (1-t) d_2\left(g, f_0\right)^2 + t d_2\left(g, f_1\right)^2 - t(1-t) d_2\left(f_0, f_1\right)^2,$$

which completes the proof. □

1.3 Equivalent conditions for CAT(0)

We first give equivalent conditions for a *complete* metric space to be CAT(0) in Theorem 1.3.2. Then in Theorem 1.3.3 we give equivalent conditions for a *geodesic* metric space to be CAT(0).

Definition 1.3.1 (4-point property). A metric space (X, d) is said to have the *4-point property* if for every $x_1, x_2, y_1, y_2 \in X$ there exist $x_1', x_2', y_1', y_2' \in \mathbb{R}^2$ such that $d(x_i, y_j) = \|x_i' - y_j'\|$ for $i, j \in \{1, 2\}$, and $d(x_1, x_2) \leq \|x_1' - x_2'\|$ and $d(y_1, y_2) \leq \|y_1' - y_2'\|$. Such a quadruple of points of $\mathbb{R}^2$ is called a *subembedding* of x_1, x_2, y_1, y_2.

Theorem 1.3.2. *Let (X, d) be a complete metric space. The following conditions are equivalent:*
(i) *The space (X, d) is CAT(0).*
(ii) *Each pair of points in X has approximate metric midpoints and X enjoys the 4-point property.*
(iii) *For every pair of points $x, y \in X$ there exists $m \in X$ such that for each $z \in X$ we have*
$$d(m, z)^2 \leq \tfrac{1}{2} d(x, z)^2 + \tfrac{1}{2} d(y, z)^2 - \tfrac{1}{4} d(x, y)^2.$$
(iv) *For every pair of points $x, y \in X$ and $\varepsilon > 0$ there exists $m \in X$ such that for each $z \in X$ we have*
$$d(m, z)^2 \leq \tfrac{1}{2} d(x, z)^2 + \tfrac{1}{2} d(y, z)^2 - \tfrac{1}{4} d(x, y)^2 + \varepsilon. \tag{1.3.13}$$

Proof. Step 1: We start with the implication (i) $\Longrightarrow$ (ii). Each pair of point has a midpoint, hence it suffices to show the 4-point property. Choose four points $x_1, y_1, x_2, y_2 \in X$ and consider the quadrilateral $Q \subset \mathbb{R}^2$ formed by comparison triangles $\Delta(\overline{x_1}, \overline{x_2}, \overline{y_1})$ and $\Delta(\overline{x_1}, \overline{x_2}, \overline{y_2})$ with a common edge $[\overline{x_1}, \overline{x_2}]$ and with $\overline{y_1}$ and $\overline{y_2}$ on opposite sides of the line containing $[\overline{x_1}, \overline{x_2}]$.

If Q is convex, then the diagonals $[\overline{x_1}, \overline{x_2}]$ and $[\overline{y_1}, \overline{y_2}]$ intersect at a point $\overline{z}$. Let $z \in [x_1, x_2]$ be the point such that $d(x_1, z) = \|\overline{x_1} - \overline{z}\|$. By the definition of CAT(0) we get
$$d(y_1, y_2) \leq d(y_1, z) + d(z, y_2) \leq \|\overline{y_1} - \overline{z}\| + \|\overline{z} - \overline{y_2}\| = \|\overline{y_1} - \overline{y_2}\|,$$
which shows that the vertices of Q form a subembedding of x_1, y_1, x_2, y_2.

If Q is not convex, then an easy application of Alexandrov's lemma (Exercise 1.3) yields a quadruple of points of $\mathbb{R}^2$ satisfying the condition of the 4-point property.

Step 2: To prove the implication (ii) $\Longrightarrow$ (iii), we first show that (ii) ensures the existence of metric midpoints. Given a pair of points $x, y \in X$, choose a sequence of approximate metric midpoints $(m_i)_{i \in \mathbb{N}}$ such that
$$\max\{d(x, m_i), d(y, m_i)\} \leq \tfrac{1}{2} d(x, y) + \tfrac{1}{i}.$$

We need to show that this sequence is Cauchy. Let (x', m_i', y, m_j') be a subembedding of (x, m_i, y, m_j). Then

$$d(x,y) \leq \|x' - y'\| \leq \|x' - m_i'\| + \|m_i' - y'\| = d(x, m_i) + d(m_i, y) \\ \leq d(x,y) + \tfrac{2}{i},$$

which implies that the sequence (m_i') converges to the metric midpoint of $[x', y']$. Since $d(m_i, m_j) \leq \|m_i' - m_j'\|$, we get that (m_i) is Cauchy, and hence converges to a point $m \in X$, which is clearly a metric midpoint of $[x, y]$.

It remains to show that for each $z \in X$, the inequality in (iii) holds. Find a subembedding $(z', x', m', y') \subset \mathbb{R}^2$ of (z, x, m, y). Since $d(z, m) \leq \|z' - m'\|$, we get the inequality in (iii) by an elementary calculation in $\mathbb{R}^2$.

Step 3: We prove (iii) $\Longrightarrow$ (i). Applying (iii) with $z := x$ and then with $z := y$ gives the existence of metric midpoints. Proposition 1.1.3 then yields that X is a geodesic space. The proof that it has the CAT(0) property is identical to that of the implications (ii) $\Longrightarrow$ (iii) and (iii) $\Longrightarrow$ (i) of Theorem 1.3.3 below.

Step 4: The implication (iii) $\Longrightarrow$ (iv) is trivial. We will prove its converse. For each $\varepsilon > 0$, the condition in (iv) gives a point m_ε satisfying (1.3.13). If we apply (1.3.13) first with $z := x$ and then with $z := y$, and add the resulting inequalities up, we obtain

$$\tfrac{1}{2}d(m_\varepsilon, x)^2 + \tfrac{1}{2}d(m_\varepsilon, y)^2 \leq \tfrac{1}{4}d(x,y)^2 + \varepsilon.$$

On the other hand, invoking the condition in (iv) with $\delta > 0$ and $z := m_\varepsilon$ yields

$$d(m_\delta, m_\varepsilon)^2 \leq \tfrac{1}{2}d(x, m_\varepsilon)^2 + \tfrac{1}{2}d(y, m_\varepsilon)^2 - \tfrac{1}{4}d(x,y)^2 + \delta \leq \varepsilon + \delta,$$

which shows that (m_ε) is Cauchy and one can readily verify that its limit point is the desired point m in (iii). □

Using the condition in Theorem 1.3.2 (iii) as a definition of a Hadamard space is probably most elegant. One then shows the existence of geodesics as a consequence of the definition. We however choose to use the classical definition from comparison geometry.

We will now state the second theorem which gives equivalent conditions for a geodesic space to be CAT(0).

Theorem 1.3.3. *Let (X, d) be a geodesic space. The following assertions are equivalent:*
(i) *The space (X, d) is CAT(0).*
(ii) *For every $x, y, z \in X$ we have*

$$d(x,m)^2 \leq \tfrac{1}{2}d(x,y)^2 + \tfrac{1}{2}d(x,z)^2 - \tfrac{1}{4}d(y,z)^2, \tag{1.3.14}$$

where m is the midpoint of $[y, z]$.
(iii) *For every geodesic $x\colon [0,1] \to X$ and every point $p \in X$ we have*

$$d(p, x_t)^2 \leq (1-t)d(p, x_0)^2 + td(p, x_1)^2 - t(1-t)d(x_0, x_1)^2.$$

(iv) *The angle between the sides of every geodesic triangle in* X *is no greater than the angle between the corresponding sides of its comparison triangle.*

(v) *For every* $x, y, u, v \in X$ *we have*

$$d(x,u)^2 + d(y,v)^2 \leq d(x,y)^2 + d(u,v)^2 + 2d(x,v)d(y,u)\,. \tag{1.3.15}$$

(vi) *For every* $x, y, u, v \in X$ *we have*

$$d(x,u)^2 + d(y,v)^2 \leq d(x,y)^2 + d(y,u)^2 + d(u,v)^2 + d(v,x)^2\,. \tag{1.3.16}$$

(vii) *The space* (X,d) *is Busemann, and*

$$d(x,y)d(u,v) \leq d(x,u)d(y,v) + d(x,v)d(y,u)\,, \tag{1.3.17}$$

for all $x, y, u, v \in X$.

Proof. The implication (i) $\Longrightarrow$ (ii) follows from (1.2.3). To see that (ii) implies (iii) it suffices to show the inequality in (iii) for all dyadic $t \in [0,1]$. We will proceed by induction. The inequality obviously holds for $t = 0$ and $t = 1$. Assume it holds for each $t = k/2^n$, where $k = 1, \ldots, 2^n$ and $n \in \mathbb{N}$ is fixed. We want to show it holds also for $t = k/2^{n+1}$, where $k = 1, \ldots, 2^{n+1}$. For k even it is just the assumption. Fix $t := k/2^{n+1}$ with k odd and put $s := 2^{-n-1}$. By (ii) we have

$$d\left(p, x_t\right)^2 \leq \tfrac{1}{2} d\left(p, x_{t-s}\right)^2 + \tfrac{1}{2} d\left(p, x_{t+s}\right)^2 - \tfrac{1}{4} d\left(x_{t-s}, x_{t+s}\right)^2,$$

and by the assumption

$$d\left(p, x_{t\pm s}\right)^2 \leq (1 - t \mp s) d\left(p, x_0\right)^2 + (t \pm s) d\left(p, x_1\right)^2 - (1 - t \mp s)(t \pm s) d\left(x_0, x_1\right)^2,$$

which together gives

$$\begin{aligned} d\left(p, x_t\right)^2 &\leq (1-t) d\left(p, x_0\right)^2 + t d\left(p, x_1\right)^2 \\ &\quad - \left[s^2 + \tfrac{1}{2}(1 - t - s)(t + s) + \tfrac{1}{2}(1 - t + s)(t - s)\right] d\left(x_0, x_1\right)^2 \\ &= (1-t) d\left(p, x_0\right)^2 + t d\left(p, x_0\right)^2 - t(1-t) d\left(x_0, x_1\right)^2\,. \end{aligned}$$

This shows (iii).

To prove (iii) $\Longrightarrow$ (i) we again choose a geodesic triangle $\Delta(p,q,r) \subset X$ along with $x \in [p,q]$ and $y \in [p,r]$. Let $\overline{\overline{y}}$ denote the corresponding point for y in the comparison triangle $\Delta(\overline{\overline{p}}, \overline{\overline{x}}, \overline{\overline{r}})$, and let $\overline{x}$ and $\overline{y}$ be the comparison points for x and y, respectively, in the comparison triangle $\Delta(\overline{p}, \overline{q}, \overline{r})$. By (iii) we have $d(x,y) \leq \|\overline{\overline{x}} - \overline{\overline{y}}\|$ and $d(x,r) \leq \|\overline{x} - \overline{r}\|$. These two inequalities give $\overline{\alpha}(x,p,y) \leq \overline{\alpha}(x,p,r)$ and $\overline{\alpha}(x,p,r) \leq \overline{\alpha}(q,p,r)$, respectively, which together yields $\overline{\alpha}(x,p,y) \leq \overline{\alpha}(q,p,r)$. Therefore $d(x,y) \leq \|\overline{x} - \overline{y}\|$.

The implications (i) $\Longrightarrow$ (iv) follows from the definition of an angle.

Let us next show (iv) $\Longrightarrow$ (iii). Recall that (iii) is equivalent to (1.2.1). Choose a geodesic triangle $\Delta(p,q,r)$ and a point $x \in [q,r]$. Fix a geodesic $[p,x]$ and let

$\triangle(\overline{p}, \overline{q}, \overline{x})$ and $\triangle(\overline{p}, \overline{r}, \overline{x})$ be comparison triangles for $\triangle(p, q, x)$ and $\triangle(p, r, x)$, respectively, with a common edge $[\overline{p}, \overline{x}]$ and $\overline{q}$ and $\overline{r}$ on opposite sides of the line containing $[\overline{p}, \overline{x}]$. Since by (iv) we have $\alpha(p, x, r) \le \overline{\alpha}(p, x, r)$ and $\alpha(p, x, q) \le \overline{\alpha}(p, x, q)$, and also $\alpha(q, x, r) = \pi$, we get that $\overline{\alpha}(p, x, r) + \overline{\alpha}(p, x, q) \ge \pi$. Applying Alexandrov's lemma (Exercise 1.3) yields that

$$d(p, x) \le \left\| \overline{\overline{p}} - \overline{\overline{x}} \right\|,$$

where $\overline{\overline{x}}$ is the comparison point for x in the comparison triangle $\triangle(\overline{\overline{p}}, \overline{\overline{q}}, \overline{\overline{r}})$ for the geodesic triangle $\triangle(p, q, r)$.

The implications (i) $\Longrightarrow$ (v) $\Longrightarrow$ (vi) follow immediately from Corollary 1.2.5.

We will show (vi) implies (ii) in several steps. Step 1: Inequality (1.3.16) forces X to be uniquely geodesic, for if $\gamma, \eta\colon [0, 1] \to X$ were two geodesics connecting a pair of points $x, y \in X$, applying (1.3.16) to $x, \gamma_{\frac{1}{2}}, y, \eta_{\frac{1}{2}}$ yields $\gamma_{\frac{1}{2}} = \eta_{\frac{1}{2}}$. Hence $\gamma_{m/2^n} = \eta_{m/2^n}$ for every $n \in \mathbb{N}$ and $m = 1, \ldots, 2^n$, which finally gives $\gamma = \eta$.

Step 2: For a triple $x, y, z \in X$ denote y' the midpoint of $[x, y]$ and z' the midpoint of $[x, z]$. We claim that

$$d\left(y', z'\right) \le d(y, z)\,. \tag{1.3.18}$$

Indeed, applying (1.3.16) to x, y', y, z' yields

$$d(x, y)^2 + d\left(y', z'\right)^2 \le \tfrac{1}{2} d(x, y)^2 + d\left(y, z'\right)^2 + \tfrac{1}{4} d(x, z)^2\,,$$

and similarly we obtain

$$d(x, z)^2 + d\left(y', z'\right)^2 \le \tfrac{1}{2} d(x, z)^2 + d\left(y', z\right)^2 + \tfrac{1}{4} d(x, y)^2\,,$$

which together gives

$$2d\left(y', z'\right)^2 + \tfrac{1}{4} d(x, y)^2 + \tfrac{1}{4} d(x, z)^2 \le d\left(y, z'\right)^2 + d\left(y', z\right)^2\,. \tag{1.3.19}$$

We next apply (1.3.16) to y, z, z', y' and get

$$d\left(y, z'\right)^2 + d\left(y', z\right)^2 \le d(y, z)^2 + \tfrac{1}{4} d(x, y)^2 + \tfrac{1}{4} d(x, z)^2 + d\left(y', z'\right)^2\,. \tag{1.3.20}$$

Combining (1.3.19) and (1.3.20) finishes the claim.

Step 3: For every $w, x, y, z \in X$ we have

$$d(w, y)^2 + d(x, z)^2 \le 2d(w, x)^2 + d(x, y)^2 + \tfrac{1}{2} d(y, z)^2 + d(z, w)^2\,. \tag{1.3.21}$$

Indeed, let v be the midpoint of $[y, z]$ and apply (1.3.16) to w, x, y, v and then to w, x, v, z, which already gives (1.3.21).

Step 4: For every $x, y, z \in X$ we have

$$\begin{aligned} &2d(x, y)^2 + 2d(x, z)^2 - d(y, z)^2 - 4d\left(x, \frac{y + z}{2}\right)^2 \\ &\ge 2\left[2d\left(\frac{x + y}{2}, y\right)^2 + 2d\left(\frac{x + y}{2}, z\right)^2 - d(y, z)^2 - 4d\left(\frac{x + y}{2}, \frac{y + z}{2}\right)^2\right]. \end{aligned} \tag{1.3.22}$$

Indeed, applying (1.3.21) to $\frac{y+z}{2}, \frac{x+y}{2}, x, z$ gives the desired inequality (1.3.22).

Step 5: Given $x, y, z \in X$, put $w := \frac{y+z}{2}$ and $y_n := (1 - \frac{1}{2^n})y + \frac{1}{2^n}x$, for every $n \in \mathbb{N}$. We want to show

$$2d(x,y)^2 + 2d(x,z)^2 - d(y,z)^2 - 4d(x,w)^2 \geq 0\,, \tag{1.3.23}$$

which already gives (vi). We first apply (1.3.22) repeatedly to get

$$\begin{aligned} &2d(x,y)^2 + 2d(x,z)^2 - d(y,z)^2 - 4d\,(x,w)^2 \\ &\geq 2^n \left[2d\,(y_n, y)^2 + 2d\,(y_n, z)^2 - d(y,z)^2 - 4d\,(y_n, w)^2\right]. \end{aligned} \tag{1.3.24}$$

Then let $v_n := \frac{1}{2}y_n + \frac{1}{2}z$ and apply (1.3.22) to the right hand side of (1.3.24) to get

$$\begin{aligned} &2d(y_n, y)^2 + 2d(y_n, z)^2 - d(y,z)^2 - 4d\,(y_n, w)^2 \\ &\geq 2\left[2d\,(v_n, y)^2 + 2d\,(v_n, z)^2 - d(y,z)^2 - 4d\,(v_n, w)^2\right]. \end{aligned} \tag{1.3.25}$$

Inequality (1.3.16) applied to v_n, y, w, z gives

$$d(v_n, w)^2 + d(y,z)^2 \leq d(v_n, y)^2 + d(v_n, z)^2 + \tfrac{1}{2}d(y,z)^2.$$

The last inequality along with (1.3.18), (1.3.24) and (1.3.25) yields

$$\begin{aligned} &2d(x,y)^2 + 2d(x,z)^2 - d(y,z)^2 - 4d(x,w)^2 \\ &\geq -2^{n+2}d\,(v_n, w)^2 \geq -2^{n+2}d\,(y_n, y)^2 = -2^{2-n}d(x,y)^2\,. \end{aligned}$$

Taking the limit $n \to \infty$ gives (1.3.23), and completes the proof (vi) $\Longrightarrow$ (ii).

It remains to show that (i) $\Longleftrightarrow$ (vii). Inequality (1.3.17) holds for every four points in $\mathbb{R}^2$ according to Remark 1.3.4. Since a CAT(0) space has the 4-point property (see Definition 1.3.1) by Theorem 1.3.2 (completeness is not needed), we immediately get that (i) $\Longrightarrow$ (vii). For the converse implication, the reader is referred to [86, Theorem 1.3]. □

A consequence of Theorem 1.3.3 (iv) is that the sum of the angles in a geodesic triangle in a Hadamard space is less or equal π.

Remark 1.3.4 (Ptolemy inequality). The classical theorem of Ptolemy states that for every four points $a, b, c, d \in \mathbb{R}^2$ we have that

$$\|a-b\|\|c-d\| \leq \|a-c\|\|b-d\| + \|a-d\|\|b-c\|,$$

and the equality holds if and only if all the four points lie on a common circle. Inequality (1.3.17) is called *Ptolemy* and metric spaces where it holds true are called *Ptolemaic*. Hence a geodesic metric space is CAT(0) if and only if it is Busemann and Ptolemaic.

Part (vi) of Theorem 1.3.3 also deserves more attention. It is quite a remarkable characterization of CAT(0) in terms of inequalities between mutual distances of finitely many points which was asked for by M. Gromov [97, Section 1.19+]. Inequality (1.3.16) holds for instance for the metric space $(Y, \sigma^{1/2})$ where (Y, σ) is an arbitrary metric space, and for ultrametric spaces; see [182, Example 1.2]. There is moreover a deeper structural meaning (1.3.16), which will be stated in Corollary 1.3.6 below.

Definition 1.3.5 (Enflo type). A metric space (X, d) has *Enflo type* $p \geq 0$ if p is the greatest number for which there is a constant $K \geq 1$ such that for every $N \in \mathbb{N}$ and $\{x_\varepsilon\}_{\varepsilon\in\{-1,1\}^N} \subset X$ we have

$$\sum_{\varepsilon\in\{-1,1\}^N} d\left(x_\varepsilon, x_{-\varepsilon}\right)^p \leq K^p \sum_{\varepsilon\sim\varepsilon'} d\left(x_\varepsilon, x_{\varepsilon'}\right)^p , \tag{1.3.26}$$

where $\varepsilon = (\varepsilon_1, \ldots, \varepsilon_N)$ and $\varepsilon \sim \varepsilon'$ stands for $\sum_1^N |\varepsilon_i - \varepsilon_i'| = 2$. The least such a constant K will be denoted by $E_p(X)$.

All metric spaces have Enflo type ≥ 1 and metric spaces with approximate metric midpoints have Enflo type ≤ 2. The reader is invited to prove these two facts in Exercise 1.11. A Ptolemaic metric space (X, d) has Enflo type 2 with $E_2(X) = \sqrt{3}$; see [154, Proposition 5.3]. As an easy corollary of Theorem 1.3.3 (vi) we obtain the Enflo type for CAT(0) spaces.

Corollary 1.3.6. *Let (X, d) be a CAT(0) space. Then it has Enflo type 2 with $E_2(X) = 1$.*

Proof. Since a CAT(0) space has (approximate metric) midpoints we know that the Enflo type is ≤ 2 by Exercise 1.11. We will prove the converse inequality. Notice that inequality (1.3.26) holds with $N = 2$ by Theorem 1.3.3 (vi). Let us assume it holds for some fixed $N \in \mathbb{N}$ and prove it holds for $N + 1$. We thus want to estimate

$$\sum_{\eta\in\{-1,1\}^{N+1}} d\left(x_\eta, x_{-\eta}\right)^2 = \sum_{\varepsilon\in\{-1,1\}^N} \left[d\left(x_{(\varepsilon,1)}, x_{(-\varepsilon,-1)}\right)^2 + d\left(x_{(\varepsilon,-1)}, x_{(-\varepsilon,1)}\right)^2\right],$$

using Theorem 1.3.3 (vi) gives

$$\begin{aligned} &\leq \sum_{\varepsilon\in\{-1,1\}^N} \Big[d\left(x_{(\varepsilon,1)}, x_{(\varepsilon,-1)}\right)^2 + d\left(x_{(-\varepsilon,-1)}, x_{(-\varepsilon,1)}\right)^2 \\ &\qquad + d\left(x_{(\varepsilon,1)}, x_{(-\varepsilon,1)}\right)^2 + d\left(x_{(\varepsilon,-1)}, x_{(-\varepsilon,-1)}\right)^2\Big] . \end{aligned}$$

By the assumption that the conclusion holds for N we obtain

$$\begin{aligned} &\leq \sum_{\varepsilon\in\{-1,1\}^N} \left[d\left(x_{(\varepsilon,1)}, x_{(\varepsilon,-1)}\right)^2 + d\left(x_{(-\varepsilon,-1)}, x_{(-\varepsilon,1)}\right)^2\right] \\ &\qquad + \sum_{\varepsilon\sim\varepsilon'} \left[d\left(x_{(\varepsilon,1)}, x_{(\varepsilon',1)}\right)^2 + d\left(x_{(\varepsilon,-1)}, x_{(\varepsilon',-1)}\right)^2\right], \end{aligned}$$

which is precisely,

$$= \sum_{\eta \sim \eta'} d\left(x_\eta, x_{\eta'}\right)^2 ,$$

and the induction step is complete. □

Finally, let us remark that yet another equivalent condition for CAT(0) in terms of probability measures is given in Theorem 2.3.5.

Exercises

Exercise 1.1. Let (X, d) be a metric space. Show that for every $x, y, z \in X$ the following inequality holds

$$\tfrac{1}{2} d(x,y)^2 \le d(x,z)^2 + d(y,z)^2 , \tag{1.3.27}$$

with the equality occurring if and only if z is a midpoint of x and y.
Hint. For every $a, b, c \in \mathbb{R}$ we have

$$0 \le (a+b-2c)^2$$
$$(a-b)^2 \le 2a^2 + 2b^2 + 4c^2 - 4ac - 4bc = 2(a-c)^2 + 2(b-c)^2 .$$

Conclude that then (1.3.27) holds in $\mathbb{R}^2$ and hence also in any other metric space. □

Exercise 1.2. Let (X, d) be a metric space. Show that for every $x, y, z \in X$ and $t \in (0, 1)$ the following inequality holds

$$t(1-t) d(x,y)^2 \le (1-t) d(x,z)^2 + t d(y,z)^2 . \tag{1.3.28}$$

The equality occurs if and only if $d(x,z) = t d(x,y)$ and $d(z,y) = (1-t) d(x,y)$.
Hint. Use the hint in Exercise 1.1 and modify as appropriate. □

Exercise 1.3 (Alexandrov's lemma). Prove the following result in $\mathbb{R}^2$, which is called Alexandrov's lemma.

Consider four points $a, b, b', c \in \mathbb{R}^2$ and assume that b and b' lie on the opposite sides of the line containing $[a, c]$. Denote α, β, γ, and α', β', γ' the angles at vertices of the triangle $\triangle(a, b, c)$, and $\triangle(a, b', c)$, respectively. Assume $\gamma + \gamma' \ge \pi$. Then

$$\|b-c\| + \left\|c-b'\right\| \le \|b-a\| + \left\|a-b'\right\| ,$$

and hence there exists a triangle $\triangle(\overline{a}, \overline{b}, \overline{b}')$ such that $\|\overline{a}-\overline{b}\| = \|a-b\|$, and $\|\overline{a}-\overline{b}'\| = \|a-b'\|$, and $\|\overline{b}-\overline{b}'\| = \|b-c\|+\|c-b'\|$. Further, if we denote $\overline{c} \in [\overline{b}, \overline{b}']$ the point with $\|b-c\| = \|\overline{b}-\overline{c}\|$, and $\overline{\alpha}, \overline{\beta}, \overline{\beta}'$ the angles at vertices $\overline{a}, \overline{b}, \overline{b}'$ in the triangle $\triangle(\overline{a}, \overline{b}, \overline{b}')$, respectively, then we have $\alpha + \alpha' \le \overline{\alpha}$, and $\beta \le \overline{\beta}$, and $\beta' \le \overline{\beta}'$.

Exercise 1.4. Show that a Banach space is CAT(0) if and only it is Hilbert.

Exercise 1.5. Prove Proposition 1.1.2 by modifying the proof of Proposition 1.1.3.

Exercise 1.6. Give an example of
(i) a metric space, which is not a length space,
(ii) a length space, which is not geodesic,
(iii) a geodesic space, which is not uniquely geodesic,
(iv) a uniquely geodesic space, which is not CAT(0).

Exercise 1.7. Show that (1.2.1) and (1.2.2) are equivalent.

Exercise 1.8. Show that a product of Hadamard spaces is a Hadamard space.

Exercise 1.9. Let $(\mathcal{H}, d)$ be a Hadamard space. Show that for every four points $x_1, x_2, x_3, x_4 \in \mathcal{H}$ we have

$$d(x_1, x_3)^2 + d(x_2, x_4)^2 \le d(x_1, x_2)^2 + d(x_2, x_3)^2 + d(x_3, x_4)^2 + d(x_4, x_1)^2 - d(m_1, m_2)^2 , \qquad (1.3.29)$$

where m_1, m_2 are the midpoints of $[x_1, x_3]$ and $[x_1, x_3]$, respectively. Note that we have equality in (1.3.29) if the space is Euclidean. Also, in the Euclidean case, the points x_1, x_2, x_3, x_4 form a parallelogram in $\mathbb{R}^2$ if and only if $m_1 = m_2$. Then (1.3.29) becomes the parallelogram identity.
Hint. Apply (1.2.3) three times: to the point x_2 and geodesic $[x_1, x_3]$, to the point x_4 and geodesic $[x_1, x_3]$, and finally to the point m_1 and geodesic $[x_2, x_4]$. □

Exercise 1.10. Prove Lemma 1.2.7.
Hint. See [51, Proposition I.1.14]. □

Exercise 1.11 (Enflo type). Prove that all metric spaces have Enflo type ≥ 1 and metric spaces with approximate metric midpoints have Enflo type ≤ 2.

Bibliographical remarks

Much of the material of this chapter has already been covered in the authoritative monograph [51] by M. Bridson and A. Haefliger and the interested reader is referred therein for further details on the geometry of Hadamard spaces. The analytical approach to nonpositive curvature was promoted by K.-T. Sturm [195]. For more details on the geometry of Hadamard spaces, we recommend also the books [3, 25, 56, 74, 104] and the survey paper [38].

The notion of a metric midpoint goes back to K. Menger [142]. Propositions 1.1.2 and 1.1.3 in this form appeared in [195, Remark 1.3] and [195, Proposition 1.2], respectively.

H. Busemann presented his approach to nonpositive curvature in geodesic spaces in [57, 58]. For a modern account on Busemann spaces, see [104] and [162]. The property in Definition 1.1.4 was used in spaces with unique geodesic *lines* to define hyperbolic spaces in the sense of Reich–Shafrir [173].

The definition of nonpositive curvature in the sense of Alexandrov comes from [2]. An overview of the historical developments before and after Alexandrov can be found in [51]. Following M. Gromov [95], we use the term CAT(0) spaces, where CAT is an acronym for Cartan–Alexandrov–Toponogov. Another name under which these spaces appear in the literature is *globally NPC spaces*, or just *NPC spaces*. There exist CAT(κ) spaces for any $\kappa \in \mathbb{R}$. The real parameter κ denotes the upper curvature bound and in the manifold case coincides with the upper bound of *sectional* curvature. If a space is of class CAT(κ), then it is also of class CAT(κ') for each $\kappa' \le \kappa$; see [51, Theorem II.1.12].

A. D. Alexandrov [2] was also the first to use an angle in geodesic spaces and therefore it is sometimes referred to as an *Alexandrov angle*. Lemma 1.2.7 is due to Alexandrov [2] as well; see also [51, Proposition I.1.14]. Proposition 1.2.8 comes from [51, Proposition II.3.3].

Alexandrov's lemma of Exercise 1.3 first appeared in [6] and has become a basic tool in comparison geometry [51, 56].

Lemma 1.2.2 comes from [51, Proposition II.1.4], we follow an alternative proof from [195, Proposition 2.3]. The statement holds true in Busemann spaces; see [104, Corollary 2.2.5].

The quadrilateral comparison in Proposition 1.2.4 goes back to J. Reshetnyak [179] and was later explored by others [51, 104, 121, 195]. Our proof follows [104, Theorem 2.3.1].

Almost all examples and constructions of Hadamard spaces presented in this chapter have already appeared in [51] and we refer the reader therein. There exist monographs exclusively on Hadamard manifolds, for instance [26, 74]. As references on the Hilbert ball, we recommend [92, 173]. The proof that the Hilbert ball is CAT(0) is scattered in [92, Lemma 17.1], [186, Lemma 2.2] and [187, Lemma 2.3]. Theorem 1.2.14 comes from M. Bridson's thesis [50], where it appeared in a more general form. Warped products were studied in [5] and in a special case in [59]. Tangent cones are described in detail in [51]. Theorem 1.2.17 is due to I. Nikolaev [153]. Proposition 1.2.18 comes from [192], but nonlinear Lebesgue spaces had been already studied by N. Korevaar and R. Schoen [121] and J. Jost [102, 104, 105].

M. Bridson's thesis [50] is the first place where various definitions of CAT(0) were proved carefully to be equivalent – after M. Gromov had asserted them to be in his essay on hyperbolic groups [95]. These equivalences were then presented in the Bridson–Haefliger book [51].

The 4-point condition first appeared in [179]. Its equivalence with CAT(0) in Theorem 1.3.2 is taken from [51, Proposition II.1.11]. For the characterization in Theorem 1.3.2 (iii) we follow [195, Proposition 2.3]. The condition in Theorem 1.3.2 (iii) is due to F. Bruhat and J. Tits [54] and is sometimes referred to as the *Bruhat–Tits in-*

equality, or the *CN-inequality*, where CN is an acronym for *courbure négative*. Part (iv) of Theorem 1.3.2 is [195, Remark 2.2].

Theorem 1.3.3 partly appeared already in [51, Proposition II.1.7]. The proof of (ii) $\Longrightarrow$ (iii) comes from [195, Proposition 2.3]. The implication (vi) $\Longrightarrow$ (ii) is a deep result due to I. Berg and I. Nikolaev [39, Theorem 6]. We follow a newer proof by T. Sato [182].

The characterization in Theorem 1.3.3 (vii) comes from [86, Theorem 1.3]. The Ptolemy inequality is equivalent to the fact that the inversion of the space is a metric space [55].

The Enflo type was introduced by P. Enflo [76–78] under the name *roundness* and belongs to the classical parts of Banach space theory [37]. Various metric types and cotypes have recently become very popular; see for instance [140, 141] for results connected to Hadamard spaces. There is also a connection with the Kirszbraun extension theorem [127].

2 Convex sets and convex functions

In this chapter we introduce convex sets and functions in Hadamard spaces. Convexity is ubiquitous in Hadamard space geometry and it will also be a central element in all our developments in the present book. Since our aim is to show that Hadamard spaces are a suitable framework for convex analysis and optimization, we focus on *analytical* aspects of convexity. After introducing basic objects and their properties in this chapter, we will get to more advanced topics in the remainder of the book.

Let $(\mathcal{H}, d)$ be a Hadamard space. Recall that a set $C \subset \mathcal{H}$ is *convex* if, given $x, y \in C$, we have $[x, y] \subset C$. We say that a function $f : \mathcal{H} \to (-\infty, \infty]$ is *convex* if, for each geodesic $\gamma : [0, 1] \to \mathcal{H}$, the function $f \circ \gamma$ is convex, that is,

$$f(\gamma_t) \leq (1-t) f(\gamma_0) + t f(\gamma_1),$$

for every $t \in (0, 1)$. We say that f is *strictly convex* if the inequality is strict whenever $\gamma_0 \neq \gamma_1$.

For instance, given a point $x_0 \in \mathcal{H}$, the function $d(\cdot, x_0)$ is convex by (1.2.4), and consequently the function $d(\cdot, x_0)^2$ is strictly convex.

2.1 Convex sets

We will first focus on (closed) convex sets. After presenting a few examples, we shall examine properties of convex sets. It should be clear that the examples below can be combined using any set operations which preserve convexity. For instance, it is easy to verify that an intersection of an arbitrarily big family of convex closed sets is itself a closed convex set.

Examples

Here are several examples of closed convex sets in a Hadamard space $(\mathcal{H}, d)$.

Example 2.1.1 (Metric balls). Let $z \in \mathcal{H}$ and $r > 0$. Then the metric ball centered at z with radius r, that is,

$$B(z, r) := \{x \in \mathcal{H} : d(x, z) \leq r\}$$

is a closed convex set by (1.2.2). The value of this example is more obvious when we recall that there exist geodesic spaces with metric balls which are not convex; see Exercise 2.1.

Example 2.1.2 (Geodesic rays and lines). An isometry $c : [0, \infty) \to \mathcal{H}$ is called a *geodesic ray* and an isometry $\gamma : \mathbb{R} \to \mathcal{H}$ is called a *geodesic line*. Then both $c([0, \infty))$

and $\gamma(\mathbb{R})$ are closed convex sets. Likewise, (the range of) a geodesic $x\colon [0,1] \to \mathcal{H}$ is a convex closed set.

Example 2.1.3 (Sublevel sets of convex functions). Let $f\colon \mathcal{H} \to (-\infty, \infty]$ be a convex lower semicontinuous function and $\beta \in \mathbb{R}$. Then the β-sublevel set of f, that is,

$$\{x \in \mathcal{H}\colon f(x) \le \beta\}$$

is a closed convex set. Note that this also captures Example 2.1.1 when we put $f := d(\cdot, z)$ and $\beta := r$.

Example 2.1.4 (Horoballs). If the function f in Example 2.1.3 is a Busemann function (see Example 2.2.10 for the definition), we call the sublevel sets *horoballs*. They carry a lot of information about the geometry of the underlying space [51]. See also Exercise 2.5.

Example 2.1.5 (Fixed point sets). Let $C \subset \mathcal{H}$ be a closed convex set and $F\colon C \to C$ be a nonexpansive mapping. Then the set of fixed points of F, that is,

$$\operatorname{Fix} F := \{x \in C\colon x = Fx\}$$

is closed and convex. According to Theorem 3.3.1 below, the set $\operatorname{Fix} F$ is nonempty, provided C is bounded.

Example 2.1.6 (Minimal displacement sets). Let $F\colon \mathcal{H} \to \mathcal{H}$ be an isometry and let

$$\delta_F(x) := d(x, Fx), \quad x \in \mathcal{H},$$

be its displacement function (see Example 2.2.9). The set

$$\operatorname{Min} F := \left\{x \in \mathcal{H}\colon \delta_F(x) = \inf_{\mathcal{H}} \delta_F\right\}$$

is then closed and convex. We say that F is *semi-simple* if $\operatorname{Min} F \neq \emptyset$. More generally, given a group G which acts by isometries on $\mathcal{H}$, we denote

$$\operatorname{Min} G := \bigcap_{F \in G} \operatorname{Min} F.$$

It is again a closed convex set.

Example 2.1.7. Let $(\mathcal{H}, d)$ be a Hadamard space and $(\mathcal{H}_\omega, d_\omega)$ be its ultrapower. Then $\mathcal{H} \subset \mathcal{H}_\omega$ is a convex closed set, which is a 1-retract.

Let $(\mathcal{H}, d)$ be a Hadamard space and $S \subset \mathcal{H}$. Then the *convex hull* of S, denoted $\operatorname{co} S$, is defined as the intersection of all convex supersets of S. And likewise the *closed convex hull* of S, denoted $\overline{\operatorname{co}}\, S$, is the intersection of all closed convex supersets of S.

Lemma 2.1.8. *Let $(\mathcal{H}, d)$ be a Hadamard space and $S \subset \mathcal{H}$. Put $C_0 := S$ and, for $n \in \mathbb{N}$, let C_n be the union of all geodesics with endpoints in C_{n-1}. Then*

$$\operatorname{co} S = \bigcup_{n=0}^{\infty} C_n.$$

Proof. It is easy to see that $\bigcup_{n=0}^{\infty} C_n$ is a convex set and contains $\operatorname{co} S$. On the other hand it is the smallest set with those properties. □

Next we define an operation closely related to convex hulls which we call a *formal convex hull*. The following notation will be useful.

Notation 2.1.9. Given $N \in \mathbb{N}$, the symbol Δ_{N-1} stands for the standard $(N-1)$-dimensional simplex in $\mathbb{R}^N$, that is, the convex hull of the canonical basis $e_1, \ldots, e_N \in \mathbb{R}^N$.

Definition 2.1.10 (Formal convex hull). Let $(\mathcal{H}, d)$ be a Hadamard space and $x_1, \ldots, x_N \in \mathcal{H}$. The *formal convex hull* of the points $x_1, \ldots, x_N$ is the smallest set $S \subset \operatorname{co}\{x_1, \ldots, x_N\} \times \Delta_{N-1}$ satisfying the following two properties:
(i) $(x_n, e_n) \in S$ for every $n = 1, \ldots, N$.
(ii) If $(u, \lambda), (w, \mu) \in S$ and $\gamma \colon [0,1] \to \mathcal{H}$ is the geodesic from u to w, then $(\gamma_t, (1-t)\lambda + t\mu) \in S$ for every $t \in [0,1]$.
The formal convex hull of the points $x_1, \ldots, x_N$ is denoted by $\operatorname{fco}\{x_1, \ldots, x_N\}$.

Note that the canonical projections

$$\pi_1 \colon \operatorname{fco}\{x_1, \ldots, x_N\} \to \operatorname{co}\{x_1, \ldots, x_N\}$$

and

$$\pi_2 \colon \operatorname{fco}\{x_1, \ldots, x_N\} \to \Delta_{N-1}$$

are both surjective. The formal convex hull will be an important tool in the proof of Theorem 4.1.2, but will be used nowhere else in this book.

Question 2.1.11. It is not known whether the following version of Mazur's theorem holds. Let $(\mathcal{H}, d)$ be a Hadamard space and $K \subset \mathcal{H}$ be compact. Is then $\overline{\operatorname{co}}\, K$ compact? This problem is probably open even for K finite.

Projections onto convex sets

It is well known that metric projections onto closed convex sets in strictly convex reflexive Banach spaces are well defined single-valued mappings, which are nonexpansive if and only if the space is Hilbert. It is comforting to learn that metric projections in Hadamard spaces behave equally well. This fact underpins a number of results in the sequel.

Given a metric space (X, d) and $S \subset X$, define the *distance function* by

$$d(x, S) := \inf_{s \in S} d(x, s), \quad x \in X.$$

Interchangeably, we use the symbol d_S for $d(\cdot, S)$. Furthermore, denote the *nearest-point mapping* by

$$P_S(x) := \{s \in S : d(x, S) = d(x, s)\}, \quad x \in X.$$

Another term for P_S is a *metric projection* onto S. Hence $P_S : X \to 2^S$ is a set-valued mapping. If the set $P_S(x)$ is a singleton, for every $x \in X$, we say that S is a *Chebyshev set*, and consider P_S as a mapping from X to S. Usually, we write $P_C x$ instead of $P_C(x)$.

The following theorem states that any convex closed subset of a Hadamard space is Chebyshev and summarizes the basic properties of the projection.

Theorem 2.1.12 (Metric projections). *Let $(\mathcal{H}, d)$ be a Hadamard space. Let $C \subset \mathcal{H}$ be a closed convex set. Then:*

(i) *The set C is Chebyshev.*

(ii) *If $x \in \mathcal{H} \setminus C$ and $z \in C \setminus \{P_C x\}$, then*

$$d(x, P_C x)^2 + d(P_C x, z)^2 \leq d(x, z)^2, \tag{2.1.1}$$

or equivalently, the angle satisfies $\alpha(x, P_C x, z) \geq \frac{\pi}{2}$.

(iii) *The mapping $P_C : \mathcal{H} \to C$ is nonexpansive, that is, we have*

$$d(P_C x, P_C y) \leq d(x, y),$$

for every $x, y \in \mathcal{H}$.

Proof. (i) Choose $x \in \mathcal{H}$. We need to show that the function $c \mapsto d(x, c)$ on C has a unique minimizer. It is clear that instead of minimizing $d(x, \cdot)$, we can equivalently minimize the function $d(x, \cdot)^2$. Since the latter is strictly convex, it has at most one minimizer. To show the existence of a minimizer, we take a minimizing sequence $(c_n) \subset C$, that is,

$$d(x, c_n)^2 \to \inf_C d(x, \cdot)^2,$$

and denote $c_{mn} := \frac{1}{2}c_m + \frac{1}{2}c_n$. Then by the convexity of C we have $c_{mn} \in C$, and (1.2.3) yields

$$d(x, c_{mn})^2 \leq \tfrac{1}{2}d(x, c_m)^2 + \tfrac{1}{2}d(x, c_n)^2 - \tfrac{1}{4}d(c_m, c_n)^2,$$

or,

$$\tfrac{1}{4}d(c_m, c_n)^2 \leq \tfrac{1}{2}d(x, c_m)^2 + \tfrac{1}{2}d(x, c_n)^2 - d(x, c_{mn})^2,$$

which, together with the fact that (c_n) is a minimizing sequence, implies that (c_n) is Cauchy. The limit point is clearly a minimizer and lies in C. To prove (ii), we denote $\gamma\colon [0,1] \to \mathcal{H}$ the geodesic connecting $P_C x$ and z. Then by (1.2.2) we obtain

$$d(x,\gamma_t)^2 \le (1-t)d(x,P_C x)^2 + td(x,z)^2 - t(1-t)d(P_C x,z)^2,$$

for any $t \in (0,1)$. Since C is convex, the geodesic γ lies in C, and

$$d(x,P_C x)^2 \le d(x,\gamma_t)^2.$$

The last two inequalities together yield

$$td(x,P_C x)^2 + t(1-t)d(P_C x,z)^2 \le td(x,z)^2.$$

After dividing by t and taking the limit $t \to 0$, we get the desired inequality

$$d(x,P_C x)^2 + d(P_C x,z)^2 \le d(x,z)^2 .$$

By the same argument, we arrive at

$$d\left(x',P_C x\right)^2 + d\left(P_C x,z'\right)^2 \le d\left(x',z'\right)^2 ,$$

for every $x' \in [x,P_C x]$ and $z' \in [z,P_C x]$, which shows $\alpha(x,P_C x,z) \ge \frac{\pi}{2}$. We finally show (iii). Corollary 1.2.5 gives

$$\begin{aligned} d(x,P_C y)^2 + d(y,P_C x)^2 \le{} & d(x,y)^2 + d(P_C x,P_C y)^2 \\ & + d(x,P_C x)^2 + d(y,P_C y)^2 . \end{aligned}$$

We apply (2.1.1) twice to arrive at

$$\begin{aligned} d(x,P_C y)^2 + d(y,P_C x)^2 \ge{} & d(x,P_C x)^2 + d(y,P_C y)^2 \\ & + 2d(P_C x,P_C y)^2 , \end{aligned}$$

and the inequality $d(P_C x,P_C y) \le d(x,y)$ follows. The proof is now complete. □

It is worth noting that the assumptions in Theorem 2.1.12 can be weakened. We need only the set C to be complete, not the whole space.

Definition 2.1.13 (Firmly nonexpansive mapping). A map $F\colon \mathcal{H} \to \mathcal{H}$ is called *firmly nonexpansive* if, for every $x, y \in \mathcal{H}$, the mapping $\Phi_{x,y}\colon [0,1] \to [0,\infty)$ defined by

$$\Phi_{x,y}(\lambda) := d\left((1-\lambda)x + \lambda Fx, (1-\lambda)y + \lambda Fy\right)$$

is nonincreasing.

In fact, metric projections are firmly nonexpansive. It follows directly by (1.2.4) and Theorem 2.1.12 (iii). Another important example of a firmly nonexpansive mapping will be the resolvent J_λ of a convex lsc function defined in (2.2.5) below, as remarked after Proposition 2.2.24. In fact, the metric projection is a special case of the resolvent; see Example 2.2.21. The resolvent R_λ of a nonexpansive mapping defined in Definition 4.2.1 is yet another instance of a firmly nonexpansive mapping; see Lemma 4.2.2.

Example 2.1.14 (Expectation of a random variable). In Section 7.3 we will use metric projections to define the conditional expectation of a random variable which takes values in a Hadamard space; see Definition 7.3.2.

Next we see that, for a Chebyshev set, nonexpansiveness of the projection in turn implies convexity of the set.

Proposition 2.1.15. *Let $(\mathcal{H}, d)$ be a Hadamard space and $C \subset \mathcal{H}$ a Chebyshev set. If P_C is nonexpansive, then C is convex.*

Proof. Choose $x, y \in C$ and $z \in [x, y]$. By the triangle inequality and the fact that P_C is nonexpansive, we have

$$d(x,z) + d(z,y) = d(x,y) \leq d\left(x, P_C z\right) + d\left(P_C z, y\right) \leq d(x,z) + d(z,y),$$

which gives $P_C z \in [x, y]$. Since P_C is nonexpansive, we conclude $d(x,z) = d(x, P_C z)$. □

At this point, we recall one of the most intriguing open problems in convex analysis: Is every Chebyshev set in a Hilbert space necessarily convex? Surprisingly, there exists a nonconvex Chebyshev set is the real hyperbolic plane $\mathbb{H}^2$.

Proposition 2.1.16. *Let $(\mathcal{H}, d)$ be a Hadamard space. If $(C_i)_{i\in I}$ is a nonincreasing family of bounded closed convex sets in $\mathcal{H}$, where I is an arbitrary directed set, then $\bigcap_{i\in I} C_i \neq \emptyset$.*

Proof. Choose $x \in \mathcal{H}$ and denote its projection onto C_i by $x_i := P_{C_i} x$. Then $(d(x, x_i))_i$ is a bounded nondecreasing net of nonnegative numbers, and hence has a limit l. If $l = 0$, then $x \in \bigcap_i C_i$. If $l > 0$, then (x_i) is Cauchy. Indeed, denote $x_{ij} := \frac{1}{2}x_i + \frac{1}{2}x_j$. By (1.2.3) we have

$$d\left(x, x_{ij}\right)^2 \leq \tfrac{1}{2} d\left(x, x_i\right)^2 + \tfrac{1}{2} d\left(x, x_j\right)^2 - \tfrac{1}{4} d\left(x_i, x_j\right)^2,$$

or,

$$\tfrac{1}{4} d\left(x_i, x_j\right)^2 \leq \tfrac{1}{2} d\left(x, x_i\right)^2 + \tfrac{1}{2} d\left(x, x_j\right)^2 - d\left(x, x_{ij}\right)^2,$$

which implies that (x_i) is Cauchy. The limit point clearly lies in $\bigcap_i C_i$. □

2.2 Convex functions

Let $(\mathcal{H}, d)$ be a Hadamard space. As it is usual in convex analysis, we work with functions which take values in $(-\infty, \infty]$. The *domain* of a function $f\colon \mathcal{H} \to (-\infty, \infty]$ is by definition the set

$$\operatorname{dom} f := \{x \in \mathcal{H} \colon f(x) < \infty\} .$$

As a convention, we usually assume $\operatorname{dom} f \neq \emptyset$ without explicit mentioning to avoid trivial cases. Likewise to our interest in *closed* convex set, we will now consider convex *lower semicontinuous* functions. Let us recall that a function $f\colon \mathcal{H} \to (-\infty, \infty]$ is *lower semicontinuous*, or *lsc*, if the set

$$\{x \in \mathcal{H} \colon f(x) \leq \alpha\}$$

is closed for each $\alpha \in \mathbb{R}$. Note that the class of convex lsc functions is stable with respect to the usual operations: If $f, g\colon \mathcal{H} \to (-\infty, \infty]$ are convex lsc, then also $f + g$ is convex lsc, and cf is convex lsc for every $c > 0$. If $f_n\colon \mathcal{H} \to (-\infty, \infty]$ is a sequence of convex lsc functions, then the functions

$$\sup_{n\in\mathbb{N}} f_n, \quad \text{and} \quad \limsup_{n\to\infty} f_n$$

are convex lsc. We also observe the following.

Lemma 2.2.1. *Let $(\mathcal{H}, d)$ be a Hadamard space and $f\colon \mathcal{H} \to (-\infty, \infty]$. Then f is convex lsc if and only its epigraph*

$$\operatorname{epi} f := \{(x, t) \in \mathcal{H} \times \mathbb{R} \colon f(x) \leq t\}$$

is a closed convex subset of $\mathcal{H} \times \mathbb{R}$.

Proof. See Exercise 2.3. ☐

A function $h\colon \mathcal{H} \to (-\infty, \infty]$ is *strongly convex* with parameter $\kappa > 0$ if,

$$h\left((1-t)x + ty\right) \leq (1-t)h(x) + th(y) - \kappa t(1-t)d(x, y)^2,$$

for every $x, y \in \mathcal{H}$ and $t \in [0, 1]$.

Remark 2.2.2. Having established this terminology, inequality (1.2.2) says that, given a fixed $z \in \mathcal{H}$, the function $d(\cdot, z)^2$ is strongly convex with parameter $\kappa = 1$. We can hence say that a geodesic metric space (X, d) is CAT(0) if (and only if) the function $x \mapsto d(x, z)^2$ is strongly convex with parameter $\kappa = 1$, for each $z \in X$.

As we would expect, strongly convex functions have 'stronger' properties; see for instance Proposition 5.1.15.

Let $(\mathcal{H}, d)$ be a Hadamard space. We say that a point $x \in \mathcal{H}$ is a *minimizer* of a function $f\colon \mathcal{H} \to (-\infty, \infty]$ if $f(x) = \inf_{\mathcal{H}} f$. The set of minimizers of f is denoted $\operatorname{Min} f$. If f is convex lsc, then $\operatorname{Min} f$ is convex and closed by Example 2.1.3.

Examples

We now present several important examples of convex lsc functions on a Hadamard space $(\mathcal{H}, d)$.

Example 2.2.3 (Indicator functions). Let $C \subset \mathcal{H}$ be a convex set. Define the *indicator function* of C by

$$\iota_C(x) := \begin{cases} 0, & \text{if } x \in C\,, \\ \infty, & \text{if } x \notin C. \end{cases}$$

Then ι_C is a convex function, and it is lsc if and only if C is closed.

Example 2.2.4 (Distance functions). Given $x_0 \in \mathcal{H}$, the function

$$x \mapsto d\left(x, x_0\right), \qquad x \in \mathcal{H}, \tag{2.2.2}$$

is convex and continuous. The function $d(\cdot, x_0)^p$ for $p > 1$ is strictly convex, and if $p = 2$, it is even strongly convex; see Remark 2.2.2. More generally, the *distance function* to a closed convex subset $C \subset \mathcal{H}$, defined as

$$d_C(x) := \inf_{c \in C} d(x, c), \qquad x \in \mathcal{H},$$

is convex and 1-Lipschitz. This follows immediately from (1.2.4) and the very definition, respectively.

Example 2.2.5. Given a finite number of points $a_1, \ldots, a_N \in \mathcal{H}$ and $(w_1, \ldots, w_N) \in \Delta_{N-1}$, we consider the function

$$f(x) := \sum_{n=1}^{N} w_n d\left(x, a_n\right)^p\,, \quad x \in \mathcal{H}\,,$$

where $p \in [1, \infty)$. The function f is convex continuous and will be used at several places in our book. In particular, we are concerned with two important cases:

(i) If $p = 1$, then f becomes the objective function in the *Fermat–Weber problem* for optimal facility location. A minimizer of f exists by Lemma 2.2.19 and is called a *median* of the points $a_1, \ldots, a_N$.

(ii) If $p = 2$, then f has a unique minimizer by Proposition 2.2.17, which is called the *mean* of the points $a_1, \ldots, a_N$. In Section 2.3 we will introduce barycenters of probability measures and observe that a mean can be viewed as the barycenter of the probability measure

$$\mu := \sum_{n=1}^{N} w_n \delta_{a_n},$$

where δ_{a_n} stands for the Dirac measure at the point a_n.

In Hilbert spaces, this definition of a mean is equivalent to the arithmetic mean, as you can prove in Exercise 2.9. See also Corollary 2.3.11 for a nice property of the mean. Medians and means have applications in Section 8.3.

Example 2.2.6. A way of generalizing the previous example is to replace the points $a_1, \dots, a_N$ by convex closed sets $C_1, \dots, C_N \subset \mathcal{H}$. Then

$$f(x) := \sum_{n=1}^{N} w_n d\left(x, C_n\right)^p, \quad x \in \mathcal{H},$$

where $p \in [1, \infty)$, is again convex and continuous.

Example 2.2.7 (Enclosing ball radius). Let $A \subset \mathcal{H}$ be a bounded set and consider the function

$$f(x) := \sup_{a \in A} d(x, a)^2, \quad x \in \mathcal{H}.$$

For a fixed $x \in \mathcal{H}$, the function value $f(x)$ represents the radius of the smallest ball enclosing the set A centered at x. A minimizer of f is therefore the center of a smallest ball enclosing A and its existence is established in Example 2.2.18.

Example 2.2.8. Let $(x_n) \subset \mathcal{H}$ be a bounded sequence. Define the function $\omega \colon \mathcal{H} \to [0, \infty)$ as

$$\omega\left(x; (x_n)\right) := \limsup_{n \to \infty} d(x, x_n)^2, \quad x \in \mathcal{H}.$$

It is locally Lipschitz, because $d(\cdot, x_n)^2$ are locally Lipschitz with a common Lipschitz constant for all $n \in \mathbb{N}$. By Lemma 2.2.15 below, it is strongly convex. The function ω will be used in the definition of the weak convergence; see (3.1.1).

Example 2.2.9 (Displacement functions). Let $F \colon \mathcal{H} \to \mathcal{H}$ be an isometry. The *displacement function* of F is the function $\delta_F \colon \mathcal{H} \to [0, \infty)$ defined by

$$\delta_F(x) := d(x, Fx), \quad x \in \mathcal{H}.$$

It is convex and 2-Lipschitz. Indeed, convexity follows from (1.2.4) and the Lipschitz property from

$$\left|d(x, Fx) - d(y, Fy)\right| \le d(x, y) + d(Fx, Fy) \le 2d(x, y).$$

Example 2.2.10 (Busemann functions). Let $c \colon [0, \infty) \to \mathcal{H}$ be a geodesic ray (see Example 2.1.2 for the definition). The function $b_c \colon \mathcal{H} \to \mathbb{R}$ defined by

$$b_c(x) := \lim_{t \to \infty} \left[d\left(x, c(t)\right) - t\right], \qquad x \in \mathcal{H},$$

is called the *Busemann function* associated to the ray c. Busemann functions are convex and 1-Lipschitz (Exercise 2.5). Concrete examples of Busemann functions are given in [51, p. 273]. Another explicit example of a Busemann function in the Hadamard space of positive definite $n \times n$ matrices with real entries can be found in [51, Proposition 10.69]. The sublevel sets of Busemann functions are called *horoballs* and were introduced in Example 2.1.4.

Example 2.2.11 (Energy functional). The energy functional is another important instance of a convex continuous function on a Hadamard space of L^2-mappings; see [192, p. 342]. Minimizers of the energy functional are called *harmonic maps* and are important in both analysis and geometry. We refer the interested reader to [96, 102, 121] and the follow-up papers. For a probabilistic approach to harmonic maps in Hadamard spaces, see [192, 193, 196].

As we shall see in the sequel, many properties of convex functions we came to expect in Hilbert spaces extend to Hadamard spaces. For an exception; see Exercise 2.4.

Lemma 2.2.12. *Let $(\mathcal{H}, d)$ be a Hadamard space and $f\colon \mathcal{H} \to (-\infty, \infty]$ be a convex lsc function. Then f is bounded from below on bounded sets.*

Proof. Let $C \subset \mathcal{H}$ be bounded, and without loss of generality assume that C is closed convex. If $\inf_C f = -\infty$, then the sets $S_k := \{x \in C\colon f(x) \le -k\}$ for $k \in \mathbb{N}$ are all nonempty. Since each S_k is closed convex and bounded, Proposition 2.1.16 yields a point $z \in \bigcap_{k\in\mathbb{N}} S_k$. Clearly $f(z) = -\infty$, which is not possible. □

Lemma 2.2.13. *Let $f\colon \mathcal{H} \to (-\infty, \infty]$ be a convex lsc function. For each $x_0 \in \mathcal{H}$ there exist constants $\alpha, \beta \in \mathbb{R}$ such that*

$$f(x) \ge \alpha + \beta d(x, x_0),$$

for every $x \in \mathcal{H}$.

Proof. Assume that this is not the case, that is, for each $k \in \mathbb{N}$, there exists $x_k \in \mathcal{H}$ such that

$$f(x_k) < -k\left[d(x_k, x_0) + 1\right].$$

Then we have

$$\liminf_{k\to\infty} f(x_k) \le -\limsup_{k\to\infty} k\left[d(x_k, x_0) + 1\right] \le -\infty,$$

which via Lemma 2.2.12 implies that (x_k) is unbounded. Choose $y \in \operatorname{dom} f$ and put

$$z_k := (1 - t_k)y + t_k x_k, \quad \text{with } t_k := \frac{1}{\sqrt{k}d(y, x_k)}.$$

Then $z_k \to y$. By convexity,

$$\begin{aligned} f(z_k) &\le (1 - t_k)f(y) + t_k f(x_k) \\ &\le (1 - t_k)f(y) - t_k k\left[d(x_k, x_0) + 1\right] \\ &\le (1 - t_k)f(y) - \sqrt{k}\,\frac{d(x_k, x_0) + 1}{d(x_k, y)}. \end{aligned}$$

Thus, by lower semicontinuity, we get

$$f(y) \le \liminf_{k\to\infty} f(z_k) \le -\infty,$$

which is not possible. □

Lemma 2.2.14. *Let $f\colon \mathcal{H} \to (-\infty, \infty]$ be a strongly convex lsc function. Then f is bounded from below.*

Proof. Let f be strongly convex with parameter $\kappa > 0$ and $(x_n) \subset \mathcal{H}$ be a minimizing sequence of f, that is, $f(x_n) \to \inf f$. By contradiction, assume $\inf f = -\infty$. Then (x_n) is unbounded by Lemma 2.2.12 and so is $y_n := \frac{1}{2}x + \frac{1}{2}x_n$ for some fixed $x \in \operatorname{dom} f$. By strong convexity

$$f(y_n) \le \frac{1}{2}f(x) + \frac{1}{2}f(x_n) - \frac{\kappa}{4}d(x, x_n)^2.$$

Lemma 2.2.13 gives $\alpha, \beta \in \mathbb{R}$ such that

$$f(y) \ge \alpha + \beta d(x, y),$$

for every $y \in \mathcal{H}$. Hence

$$\alpha + \beta d(x, x_n) \le \frac{1}{2}f(x) + \frac{1}{2}f(x_n) - \frac{\kappa}{4}d(x, x_n)^2,$$

which is not possible, so f is bounded from below. □

Lemma 2.2.15. *Let $f_n\colon \mathcal{H} \to (-\infty, \infty]$ be strongly convex functions with common parameter κ, for every $n \in \mathbb{N}$. Then the function $f := \limsup_{n\to\infty} f_n$ is strongly convex with parameter κ. Let $f_i\colon \mathcal{H} \to (-\infty, \infty]$ be strongly convex functions with common parameter κ, for every $i \in I$, where I is an arbitrary index set. Then the function $f := \sup_{i\in I} f_i$ is strongly convex with parameter κ.*

Proof. See Exercise 2.7. □

Question 2.2.16 (Lower semicontinuity of ultra-extensions). Assume $(\mathcal{H}, d)$ is a Hadamard space and $(\mathcal{H}_\omega, d_\omega)$ is an ultralimit of the constant sequence $\mathcal{H}_n := \mathcal{H}$ for each $n \in \mathbb{N}$. Given a convex lsc function $f\colon \mathcal{H} \to (-\infty, \infty]$, define its *ultra-extension* by

$$f_\omega(x_\omega) := \inf\left\{\omega\text{-}\lim_n f(x_n) : (x_n) \in x_\omega\right\}, \quad x_\omega \in \mathcal{H}_\omega.$$

Since f is bounded from below on bounded sets by Lemma 2.2.12, we have $f_\omega > -\infty$. The ultra-extension of a convex function $f\colon \mathcal{H} \to (-\infty, \infty]$ is a convex function $f_\omega\colon \mathcal{H}_\omega \to (-\infty, \infty]$ and we have $f_\omega \restriction_{\mathcal{H}} = f$. We do not know whether the ultra-extension of a convex lsc function is lsc.

Minimizers of convex functions

Next we establish conditions which ensure the existence and uniqueness of minimizers of convex lsc functions. Clearly, a strictly convex function admits at most one minimizer. Strong convexity is sufficient to guarantee also the existence as the following proposition states.

Proposition 2.2.17. *Let $(\mathcal{H}, d)$ be a Hadamard space and $f\colon \mathcal{H} \to (-\infty, \infty]$ be a lsc strongly convex function with parameter $\kappa > 0$. Then there exists a unique minimizer $x \in \mathcal{H}$ of f and each minimizing sequence converges to x. Moreover,*

$$f(x) + \kappa d(x, y)^2 \le f(y)\,, \tag{2.2.3}$$

for each $y \in \mathcal{H}$.

Proof. Let $(x_n) \subset \mathcal{H}$ be a minimizing sequence of f. Lemma 2.2.14 states that f is bounded from below. Denote x_{mn} the midpoint of $[x_m, x_n]$. Then again by strong convexity

$$f\left(x_{mn}\right) \le \frac{1}{2} f\left(x_m\right) + \frac{1}{2} f\left(x_n\right) - \frac{\kappa}{4} d\left(x_m, x_n\right)^2,$$

which yields that (x_n) is Cauchy, and hence converges to a point $x \in \mathcal{H}$. Finally, we use the lower semicontinuity of f to conclude that $x \in \operatorname{Min} f$. Uniqueness is obvious and consequently also the fact that any other minimizing sequence converges to x.

To show (2.2.3) we choose $y \in \mathcal{H}$ and denote $\gamma\colon [0, 1] \to \mathcal{H}$ the geodesic from x to y. Strong convexity gives

$$f(x) < f\left(\gamma_t\right) \le (1-t) f(x) + t f(y) - \kappa t(1-t) d(x, y)^2,$$

for every $t \in (0, 1)$. Dividing by t and taking the limit $t \to 0$ yields (2.2.3). The proof is now complete. □

Note that Theorem 2.1.12 (i) and (ii) is a special case of Proposition 2.2.17. Indeed, let $C \subset \mathcal{H}$ be a closed convex set and $x \in \mathcal{H} \setminus C$. Put

$$f(y) := \begin{cases} d(x, y)^2, & \text{if } y \in C\,, \\ \infty, & \text{on } y \in \mathcal{H} \setminus C. \end{cases}$$

Then f is strongly convex due to the nonpositive curvature of the Hadamard space.

Example 2.2.18 (Minimal enclosing ball center). Let $A \subset \mathcal{H}$ be a bounded set and consider the function from Example 2.2.7, that is,

$$f(x) := \sup_{a \in A} d(x, a)^2\,, \quad x \in \mathcal{H}.$$

According to Lemma 2.2.15, the function is strongly convex, and therefore has a unique minimizer due to Proposition 2.2.17. This minimizer is the center of the minimal ball enclosing the set A. If $A := \{a_1, \ldots, a_N\}$ is finite, finding a minimizer of f amounts to the min-max problem

$$\min_{x \in \mathcal{H}} \max_n d\left(x, a_n\right)^2,$$

where n runs over $\{1, \ldots, N\}$.

Another condition to guarantee the existence of a minimizer is coercivity.

Lemma 2.2.19. *Let $f\colon \mathcal{H} \to (-\infty, \infty]$ be a convex lsc function which satisfies $f(x) \to \infty$ whenever $d(x, x_0) \to \infty$ for some $x_0 \in \mathcal{H}$. Then f has a minimizer.*

Proof. By Lemma 2.2.12 we know that f is bounded from below on bounded sets, and hence it is bounded from below on $\mathcal{H}$ since $f(x) \to \infty$ whenever $d(x, x_0) \to \infty$. Therefore $\inf_{\mathcal{H}} f > -\infty$, and we can apply Proposition 2.1.16 to the sublevel sets

$$C_n := \left\{ x \in \mathcal{H} : f(x) \le \inf_{\mathcal{H}} f + \frac{1}{n} \right\},$$

to get the existence of a minimizer. □

Notice that the above lemma covers, as a trivial case, the situation when $\mathcal{H}$ is bounded. Then every convex lsc function is coercive and has a minimizer.

The *Moreau–Yosida envelope* of a convex lsc function $f\colon \mathcal{H} \to (-\infty, \infty]$ with parameter $\lambda > 0$ is defined as

$$f_\lambda(x) := \inf_{y \in \mathcal{H}} \left[f(y) + \frac{1}{2\lambda} d(x, y)^2 \right], \quad x \in \mathcal{H}. \tag{2.2.4}$$

Since f is convex and $d(x, \cdot)^2$ strongly convex, the function $f + d(x, \cdot)^2$ is strongly convex, and hence bounded from below (Lemma 2.2.14), which implies $f_\lambda > -\infty$. Also $f_\lambda < \infty$, unless of course $\operatorname{dom} f = \emptyset$, but we do not consider such a function. Furthermore, the function f_λ is convex, but not necessarily lsc; see Exercise 2.8. Given $\lambda, \mu > 0$, the following holds

$$(f_\lambda)_\mu (x) = f_{\lambda+\mu}(x), \quad x \in \mathcal{H},$$

as one can easily check. This gives then rise to the *Hamilton–Jacobi semigroup*.

As we already noted above, the function $f + d(x, \cdot)^2$ is strongly convex and the following mapping is therefore well defined.

Definition 2.2.20 (Resolvent of a function). Let $(\mathcal{H}, d)$ be a Hadamard space and $f\colon \mathcal{H} \to (-\infty, \infty]$ be convex lsc. For $\lambda > 0$, define the *resolvent* of f as

$$J_\lambda x := \operatorname*{arg\,min}_{y \in \mathcal{H}} \left[f(y) + \frac{1}{2\lambda} d(x, y)^2 \right], \quad x \in \mathcal{H}, \tag{2.2.5}$$

and put $J_0 x := x$ for each $x \in \mathcal{H}$.

If a confusion is likely, one should use the notation J_λ^f to specify which function a resolvent belongs to. But it will usually be clear from the context, so we omit the superscript. Some authors use the term *proximal mapping* for the resolvent and denote it P_λ instead of J_λ, but we will not do that.

Example 2.2.21 (Metric projections revisited). If f is the indicator function of a closed convex set (see Example 2.2.3 for the definition), then the resolvent is the metric projection onto C, that is, $J_\lambda = P_C$ for each $\lambda > 0$.

Theorem 2.2.22. *Let $(\mathcal{H}, d)$ be a Hadamard space and $f\colon \mathcal{H} \to (-\infty, \infty]$ be convex lsc. Given $\lambda > 0$, the resolvent J_λ of f is nonexpansive, that is,*

$$d\left(J_\lambda x_0, J_\lambda x_1\right) \le d\left(x_0, x_1\right), \quad x_0, x_1 \in \mathcal{H}.$$

Proof. Denote $y\colon [0,1] \to \mathcal{H}$ the geodesic from $J_\lambda x_0$ to $J_\lambda x_1$. By the convexity of f we have

$$f\left(y_t\right) + f\left(y_{1-t}\right) \le f\left(y_0\right) + f\left(y_1\right). \tag{2.2.6}$$

Proposition 1.2.4 gives for any $t \in (0,1)$ the following estimate

$$\begin{aligned} d\left(y_t, x_0\right)^2 + d\left(y_{1-t}, x_1\right)^2 \le\ & d\left(y_0, x_0\right)^2 + d\left(y_1, x_1\right)^2 \\ & + 2t^2 d\left(y_0, y_1\right)^2 \\ & + t d\left(x_0, x_1\right)^2 - t d\left(y_0, y_1\right)^2 \\ & - t\left[d\left(x_0, x_1\right) - d\left(y_0, y_1\right)\right]^2 . \end{aligned} \tag{2.2.7}$$

Multiplying (2.2.7) by $\frac{1}{2\lambda}$ and adding to (2.2.6) yields

$$0 \le 2t d\left(y_0, y_1\right)^2 + d\left(x_0, x_1\right)^2 - d\left(y_0, y_1\right)^2 + \left[d\left(x_0, x_1\right) - d\left(y_0, y_1\right)\right]^2 ,$$

since we have

$$f\left(y_t\right) + \frac{1}{2\lambda} d\left(y_t, x_0\right)^2 + f\left(y_{1-t}\right) + \frac{1}{2\lambda} d\left(y_{1-t}, x_1\right)^2 \ge f_\lambda\left(y_0\right) + f_\lambda\left(y_1\right)$$

by the definition of $y_0 = J_\lambda x_0$ and $y_1 = J_\lambda x_1$. Taking the limit $t \to 0$ in the second last inequality finishes the proof. □

The following lemma is similar to the variational inequality in (2.2.3), but is more subtle since we take the structure of the strongly convex function in question into account. One can also view it as a discrete version of the evolution variation inequality from Theorem 5.1.11.

Lemma 2.2.23. *Let $f\colon \mathcal{H} \to (-\infty, \infty]$ be a convex lsc function. Then for every $x, y \in \mathcal{H}$ and $\lambda > 0$, we have*

$$\frac{1}{2\lambda} d\left(J_\lambda x, y\right)^2 - \frac{1}{2\lambda} d(x,y)^2 \le f(y) - f_\lambda(x).$$

Proof. Denote $\gamma\colon [0,1] \to \mathcal{H}$ the geodesic connecting $J_\lambda x$ and y. Applying (1.2.2) and the convexity of f yields

$$\begin{aligned} f_\lambda(x) &\le f\left(\gamma_t\right) + \frac{1}{2\lambda} d\left(x, \gamma_t\right)^2 \\ &\le (1-t) f_\lambda(x) + t f(y) + \frac{t}{2\lambda} d(x,y)^2 - \frac{t(1-t)}{2\lambda} d\left(J_\lambda x, y\right)^2 , \end{aligned}$$

for any $t \in (0,1)$. After dividing by t we obtain

$$0 \le -f_\lambda(x) + f(y) + \frac{1}{2\lambda} d(x,y)^2 - \frac{1-t}{2\lambda} d\left(J_\lambda x, y\right)^2,$$

and passing to the limit $t \to 0$ finally gives the desired inequality. □

The resolvent J_λ is not a semigroup in the variable λ. The following Proposition 2.2.24 describes the dependence of J_λ on λ.

Proposition 2.2.24 (Resolvent identity). *Let $(\mathcal{H}, d)$ be a Hadamard space and $f\colon \mathcal{H} \to (-\infty, \infty]$ be convex lsc. Then the following identity holds*

$$J_\lambda x = J_\mu \left(\frac{\lambda - \mu}{\lambda} J_\lambda x + \frac{\mu}{\lambda} x \right), \quad x \in \mathcal{H},$$

for every $\lambda > \mu > 0$.

Proof. Denote $y := \frac{\lambda-\mu}{\lambda} J_\lambda x + \frac{\mu}{\lambda} x$. By the definition of the resolvent we have the inequalities

$$f\left(J_\mu(y)\right) + \frac{1}{2\mu} d\left(J_\mu y, y\right)^2 \le f\left(J_\lambda x\right) + \frac{1}{2\mu} d\left(J_\lambda x, y\right)^2, \tag{2.2.8}$$

and,

$$f\left(J_\lambda(x)\right) + \frac{1}{2\lambda} d\left(J_\lambda x, x\right)^2 \le f\left(J_\mu y\right) + \frac{1}{2\lambda} d\left(J_\mu y, x\right)^2. \tag{2.2.9}$$

Summing (2.2.8) and (2.2.9) yields

$$\frac{1}{2\mu} d\left(J_\mu y, y\right)^2 + \frac{1}{2\lambda} d\left(J_\lambda x, x\right)^2 \le \frac{1}{2\mu} d\left(J_\lambda x, y\right)^2 + \frac{1}{2\lambda} d\left(J_\mu y, x\right)^2.$$

By a simple computation we have

$$\frac{1}{\mu} d\left(J_\mu y, y\right)^2 + \frac{\lambda}{(\lambda-\mu)^2} d\left(y, x\right)^2 \le \frac{\mu}{(\lambda-\mu)^2} d\left(y, x\right)^2 + \frac{1}{\lambda} d\left(J_\mu y, x\right)^2,$$

and furthermore,

$$\frac{\lambda}{\mu} d\left(J_\mu y, y\right)^2 + \frac{\lambda}{\lambda-\mu} d\left(y, x\right)^2 \le d\left(J_\mu y, x\right)^2.$$

On the other hand, by (1.3.28) we know that

$$d\left(J_\mu y, x\right)^2 \le \frac{\lambda}{\mu} d\left(J_\mu y, y\right)^2 + \frac{\lambda}{\lambda-\mu} d\left(y, x\right)^2.$$

Hence all of the above inequalities are actually equalities. In particular, by the uniqueness of the minimizer, we get from (2.2.8) that $J_\lambda x = J_\mu y$, which finishes the proof. □

As a corollary, we obtain that a resolvent is even firmly nonexpansive; see Definition 2.1.13. Indeed, denote x_t the geodesic from x_0 to $J_\lambda x_0$, and y_t the geodesic from y_0 to $J_\lambda y_0$. The resolvent identity from Proposition 2.2.24 gives

$$d\left(J_\lambda x_0, J_\lambda y_0\right) = d\left(J_{(1-t)\lambda} x_t, J_{(1-t)\lambda} y_t\right) \leq d\left(x_t, y_t\right),$$

for every $t \in (0,1)$. The joint convexity of the metric (1.2.4) then gives that J_λ is firmly nonexpansive.

The resolvents are essential in the proof of the existence of harmonic maps. Indeed, the energy functional is convex and lsc on a suitable Hadamard space Y of L^2-mappings (Example 2.2.11) and $J_\lambda y$, with an arbitrary $y \in Y$, converges to a minimizer of the energy functional as $\lambda \to \infty$. This relies upon the following theorem. For the details, see [103–106].

Theorem 2.2.25. *Let $(\mathcal{H}, d)$ be a Hadamard space, $f\colon \mathcal{H} \to (-\infty, \infty]$ be convex lsc and $x \in \mathcal{H}$. If $(J_{\lambda_n} x)_{n\in\mathbb{N}}$ is bounded for some sequence $\lambda_n \to \infty$, then $\operatorname{Min} f \neq \emptyset$ and*

$$\lim_{\lambda\to\infty} J_\lambda x = P_{\operatorname{Min} f}(x),$$

where $P_{\operatorname{Min} f}$ stands for the metric projection onto $\operatorname{Min} f$.

Proof. For simplicity denote $y_\lambda := J_\lambda x$, with $\lambda > 0$. Observe that for each $n \in \mathbb{N}$, the point y_{λ_n} is a minimizer of the function $f + \frac{1}{2\lambda_n} d(\cdot, x)^2$ and recall that the sequence $(y_{\lambda_n})_{n\in\mathbb{N}}$ is bounded by the assumptions. These two facts imply that $(y_{\lambda_n})_{n\in\mathbb{N}}$ is a minimizing sequence of f.

Next we show that the function $\lambda \mapsto d(x, y_\lambda)$ is increasing. Suppose $0 < \mu < \lambda$. Then

$$f\left(y_\lambda\right) + \frac{1}{2\mu} d\left(x, y_\lambda\right)^2 > f\left(y_\mu\right) + \frac{1}{2\mu} d\left(x, y_\mu\right)^2,$$

and furthermore,

$$\begin{aligned} f(y_\lambda) + \frac{1}{2\lambda} d(x, y_\lambda)^2 > {} & f\left(y_\mu\right) + \frac{1}{2\lambda} d\left(x, y_\mu\right)^2 \\ & + \left(\frac{1}{2\mu} - \frac{1}{2\lambda}\right)\left[d\left(x, y_\mu\right)^2 - d\left(x, y_\lambda\right)^2\right], \end{aligned}$$

which already implies $d(x, y_\mu) < d(x, y_\lambda)$.

As the function $\lambda \mapsto d(x, y_\lambda)$ is bounded on the sequence (λ_n) and monotone, it has to be bounded on all $(0, \infty)$. Clearly, we have

$$f\left(y_\lambda\right) = \inf_{y\in\mathcal{H}} \left\{ f(y) : d(x, y) \leq d\left(x, y_\lambda\right)\right\},$$

and since $\lambda \mapsto d(x, y_\lambda)$ is increasing, we conclude that $\lambda \mapsto f(y_\lambda)$ is nonincreasing and furthermore that $f(y_\lambda)$ tends to $\inf_{\mathcal{H}} f$ as $\lambda \to \infty$.

Choose now $\varepsilon > 0$ and find $\Lambda > 0$ such that for every $\lambda \geq \mu \geq \Lambda$ the following inequality holds

$$\left| d\left(x, y_\mu\right)^2 - d\left(x, y_\lambda\right)^2 \right| < \varepsilon.$$

If $y_{\mu\lambda}$ is the midpoint of y_μ and y_λ, then we get

$$\begin{aligned} f\left(y_\mu\right) + \frac{1}{2\mu} d\left(x, y_\mu\right)^2 &\leq f\left(y_{\mu\lambda}\right) + \frac{1}{2\mu} d\left(x, y_{\mu\lambda}\right)^2 \\ &\leq f\left(y_{\mu\lambda}\right) + \frac{1}{2\mu}\left[d\left(x, y_\mu\right)^2 + \frac{\varepsilon}{2} - \frac{1}{4} d\left(y_\mu, y_\lambda\right)^2 \right] \\ &\leq f\left(y_\mu\right) + \frac{1}{2\mu}\left[d\left(x, y_\mu\right)^2 + \frac{\varepsilon}{2} - \frac{1}{4} d\left(y_\mu, y_\lambda\right)^2 \right]. \end{aligned}$$

Hence $d(y_\mu, y_\lambda)^2 < 2\varepsilon$, that is, $(y_\lambda)_{\lambda>0}$ is Cauchy. By completeness, there is a point $y \in \mathcal{H}$ such that $y_\lambda \to y$ as $\lambda \to \infty$. By the lsc of f, the point y is a minimizer of f. It is straightforward to show $y = P_{\operatorname{Min} f}(x)$. □

Proposition 2.2.26. *Let $f \colon \mathcal{H} \to (-\infty, \infty]$ be a convex lsc function and $x \in \mathcal{H}$. Then the function $\lambda \mapsto J_\lambda x$ is continuous on $(0, \infty)$ and*

$$\lim_{\lambda \to 0+} J_\lambda x = P_{\overline{\operatorname{dom} f}}(x).$$

In particular, if $x \in \overline{\operatorname{dom} f}$, the function $\lambda \mapsto J_\lambda x$ is continuous on $[0, \infty)$.

Proof. Denote $P := P_{\overline{\operatorname{dom} f}}$. The resolvent identity (Proposition 2.2.24) and Theorem 2.2.22 easily yield continuity at any positive λ. We leave the details to the reader.

To show the second part of the statement, put $x_\lambda := J_\lambda x$ and recall that by the proof of Theorem 2.2.25, we have $f(x_\lambda)$ is nonincreasing, in particular, there exists $\alpha \in \mathbb{R}$ such that $f(x_\lambda) \geq \alpha$ for any $\lambda \in (0, 1]$. Fix $\varepsilon > 0$ and find $\delta > 0$ such that

$$[d(x, Px) + \delta]^2 - d(x, Px)^2 < \varepsilon.$$

Then choose $z \in \operatorname{dom} f$ with $d(z, Px) < \delta$. Applying (2.1.1) we obtain

$$\begin{aligned} f\left(x_\lambda\right) + \frac{1}{2\lambda} d\left(x_\lambda, Px\right)^2 + \frac{1}{2\lambda} d\left(Px, x\right)^2 &\leq f\left(x_\lambda\right) + \frac{1}{2\lambda} d\left(x_\lambda, x\right)^2 \\ &\leq f(z) + \frac{1}{2\lambda} d(z, x)^2 \\ &\leq f(z) + \frac{1}{2\lambda}\left[d(x, Px) + \delta\right]^2. \end{aligned}$$

Consequently,

$$d\left(x_\lambda, Px\right)^2 \leq [d(x, Px) + \delta]^2 - d(x, Px)^2 + 2\lambda\left[f(z) - \alpha\right],$$

whenever $\lambda \in (0, 1]$. By choosing λ such that $2\lambda[f(z) - \alpha] < \varepsilon$, we finally get

$$d\left(x_\lambda, Px\right)^2 < 2\varepsilon,$$

which finishes the proof. □

The following result is due to de Giorgi.

Proposition 2.2.27. *Let $f\colon \mathcal{H} \to (-\infty, \infty]$ be convex lsc and $x \in \mathcal{H}$. Then the function $\lambda \mapsto f_\lambda(x)$ is locally Lipschitz on $(0, \infty)$ and*

$$\frac{\mathrm{d}}{\mathrm{d}\lambda} f_\lambda(x) = -\frac{d\left(x, J_\lambda x\right)^2}{2\lambda^2},$$

for almost every $\lambda \in (0, \infty)$.

Proof. Let $\mu, \lambda \in (0, \infty)$. By the definition of the resolvent we have

$$f\left(J_\lambda x\right) + \frac{1}{2\mu} d\left(J_\lambda x, x\right)^2 \geq f\left(J_\mu x\right) + \frac{1}{2\mu} d\left(J_\mu x, x\right)^2,$$

and therefore

$$\begin{aligned}\frac{\lambda - \mu}{2\lambda\mu} d\left(J_\lambda x, x\right)^2 &= \left(\frac{1}{2\mu} - \frac{1}{2\lambda}\right) d\left(J_\lambda x, x\right)^2 \\ &\geq f\left(J_\mu x\right) + \frac{1}{2\mu} d\left(J_\mu x, x\right)^2 - f\left(J_\lambda x\right) - \frac{1}{2\lambda} d\left(J_\lambda x, x\right)^2 .\end{aligned}$$

By symmetry we obtain

$$\begin{aligned}\frac{\mu - \lambda}{2\lambda\mu} d\left(J_\mu x, x\right)^2 &= \left(\frac{1}{2\lambda} - \frac{1}{2\mu}\right) d\left(J_\mu x, x\right)^2 \\ &\geq f\left(J_\lambda x\right) + \frac{1}{2\lambda} d\left(J_\lambda x, x\right)^2 - f\left(J_\mu x\right) - \frac{1}{2\mu} d\left(J_\mu x, x\right)^2 .\end{aligned}$$

The statement follows easily. □

The theory of resolvents continues in Chapter 5, where we use resolvents to define nonlinear gradient flow semigroups.

Remark 2.2.28. The existence of minimizers of the displacement function (see Example 2.2.9 for the definition) gives rise to a classification of isometries on metric spaces. Let F be an isometry on a metric space (X, d). Then it is called

- *parabolic* if d_F does not attain its infimum,
- *elliptic* if d_F attains its infimum of 0,
- *hyperbolic* if d_F attains its strictly positive infimum.

It is clear that F is semi-simple if and only if it is not parabolic; see Example 2.1.6. We refer the interested reader to the Bridson–Haefliger book [51] for more details.

Other properties of convex functions will appear at many places throughout the rest of the book. For instance, in Lemma 3.2.3 we will show that convex functions are lsc if and only they are weakly lsc.

2.3 Convexity and probability measures

Next, a relationship between convexity and probability measures is studied in detail. We derive the *analytical* properties of probability measures here and thence prepare a ground for Chapter 7, where the *stochastic* properties of probability measures will be investigated.

Let (X, d) be a metric space and denote $\mathcal{P}(X)$ the set of all probability measures on X. For $p \in [1, \infty)$ let $\mathcal{P}^p(X)$ be the set of $\mu \in \mathcal{P}(X)$ such that

$$\int_X d(x, y)^p \,\mathrm{d}\mu(y) < \infty,$$

for some, and hence all, $x \in X$. Denote $\mathcal{P}^\infty(X)$ the set of measures of $\mathcal{P}(X)$ with bounded support. Observe that $\mathcal{P}^\infty(X) \subset \mathcal{P}^p(X) \subset \mathcal{P}^1(X)$. The Dirac measure at a point $x \in X$ will be denoted δ_x.

Theorem 2.3.1 (Existence of a barycenter). *Let $(\mathcal{H}, d)$ be a Hadamard space and $\mu \in \mathcal{P}^1(\mathcal{H})$. Given $y \in \mathcal{H}$, there is a unique minimizer of the function*

$$F_y(z) := \int_{\mathcal{H}} \left[d(x, z)^2 - d(x, y)^2 \right] \mathrm{d}\mu(x), \quad z \in \mathcal{H}. \tag{2.3.10}$$

This minimizer is independent of y and is called the barycenter *of μ, denoted $b(\mu)$. Given $z \in \mathcal{H}$, the following inequality holds:*

$$\int_{\mathcal{H}} \left[d(z, x)^2 - d\left(b(\mu), x\right)^2 \right] \mathrm{d}\mu(x) \geq d\left(z, b(\mu)\right)^2. \tag{2.3.11}$$

If, moreover, $\mu \in \mathcal{P}^2(\mathcal{H})$, then

$$b(\mu) = \arg\min_{z \in \mathcal{H}} \int_{\mathcal{H}} d(x, z)^2 \,\mathrm{d}\mu(x). \tag{2.3.12}$$

Proof. First observe that the function $z \mapsto F_y(z) - F_{y'}(z)$ is constant, which means that the barycenter (if it exists) does not depend on y. Furthermore,

$$\begin{aligned} \left|F_y(z)\right| &= \left| \int_{\mathcal{H}} \left[d(x, z) - d(x, y) \right] \cdot \left[d(x, z) + d(x, y) \right] \mathrm{d}\mu(x) \right| \\ &\leq d(z, y) \cdot \left[\int_{\mathcal{H}} d(x, z) \,\mathrm{d}\mu(x) + \int_{\mathcal{H}} d(x, y) \,\mathrm{d}\mu(x) \right], \end{aligned}$$

which yields $|F_y(z)| < \infty$. We now claim that F_y is strongly convex. Indeed, let $z_t := (1-t)z_0 + tz_1$ with $t \in [0,1]$, then by (1.2.2) we have

$$\begin{aligned} F_y(z_t) &= \int_{\mathcal{H}} \left[d(x,z_t)^2 - d(x,y)^2\right] \mathrm{d}\mu(x) \\ &\le (1-t) \int_{\mathcal{H}} \left[d(x,z_0)^2 - d(x,y)^2\right] \mathrm{d}\mu(x) \\ &\quad + t \int_{\mathcal{H}} \left[d(x,z_1)^2 - d(x,y)^2\right] \mathrm{d}\mu(x) - t(1-t)d(z_0,z_1)^2 \\ &= (1-t)F_y(z_0) + tF_y(z_1) - t(1-t)d(z_0,z_1)^2 . \end{aligned} \tag{2.3.13}$$

Since F_y is obviously continuous, applying Lemma 2.2.17 gives the existence and uniqueness of the barycenter. If we now apply (2.3.13) with $z_0 := b(\mu)$, $z_1 := z$ and $y := b(\mu)$, we get

$$0 \le F_{b(\mu)}(z_t) \le tF_{b(\mu)}(z) - t(1-t)d\left(z, b(\mu)\right)^2 ,$$

and consequently for any $t \in (0,1)$ one obtains

$$\int_{\mathcal{H}} \left[d(z,x)^2 - d\left(b(\mu),x\right)^2\right] \mathrm{d}\mu(x) \ge (1-t)d\left(z, b(\mu)\right)^2 .$$

Letting $t \to 0$ yields (2.3.11). If finally $\mu \in \mathcal{P}^2(\mathcal{H})$, then we immediately get (2.3.12). □

We have already met barycenters of *discrete* measures in Example 2.2.5. The barycenter of a probability measure can be approximated via the law of large numbers in Theorem 7.2.1, which turns out to be very useful from the computational point of view; see Section 7.2.

Example 2.3.2 (Pettis integral). If H is a Hilbert space and $\mu \in \mathcal{P}^1(H)$, then

$$b(\mu) = \int_H x \, \mathrm{d}\mu(x),$$

in the weak (Pettis) sense, that is,

$$\langle b(\mu), y\rangle = \int_H \langle x, y\rangle \, \mathrm{d}\mu(x),$$

for every $y \in H$. Indeed, by the definition, $b(\mu)$ is a unique minimizer of the function

$$F\colon z \mapsto \int_H \left[d(x,z)^2 - d(x,0)^2\right] \mathrm{d}\mu(x) = \int_H \|x-z\|^2 - \|x\|^2 \, \mathrm{d}\mu(x).$$

Therefore $z = b(\mu)$ if and only if

$$\frac{\mathrm{d}}{\mathrm{d}t}F(z+ty)|_{t=0} = 2\int_H \langle z-x, y\rangle \,\mathrm{d}\mu(x) = 0,$$

for every $y \in H$.

Lemma 2.3.3. *Let $C \subset \mathcal{H}$ be a closed convex set and $\mu \in \mathcal{P}^1(\mathcal{H})$. If $\operatorname{supp}\mu \subset C$, then $b(\mu) \in C$.*

Proof. Denote $p := P_C(b(\mu))$ the projection of $b(\mu)$ onto C. Then (2.1.1) gives

$$\int_{\mathcal{H}} \left[d(p,x)^2 - d(y,x)^2\right] \mathrm{d}\mu(x) \le \int_{\mathcal{H}} \left[d\left(b(\mu),x\right)^2 - d(y,x)^2\right] \mathrm{d}\mu(x),$$

and by the definition of $b(\mu)$ we must have $b(\mu) = p$. □

Definition 2.3.4 (Variance of a measure). For $\mu \in \mathcal{P}(\mathcal{H})$ denote its *variance* by

$$\operatorname{var}\mu := \inf_{z\in\mathcal{H}} \int_{\mathcal{H}} d(x,z)^2 \,\mathrm{d}\mu(x).$$

Hence $\mathcal{P}^2(\mathcal{H})$ is the set of the probability measures of finite variance.

Let $\mu \in \mathcal{P}^2(\mathcal{H})$. Inequality (2.3.11) reads

$$\int_{\mathcal{H}} d(z,x)^2 \,\mathrm{d}\mu(x) \ge d\left(z, b(\mu)\right)^2 + \operatorname{var}\mu\,, \tag{2.3.14}$$

for every $z \in \mathcal{H}$. We call (2.3.14) the *variance inequality* for μ.

We will now give conditions on a complete metric space in terms of probability measures each of which is equivalent to CAT(0).

Theorem 2.3.5. *Let (X,d) be a complete metric space. Then the following properties are equivalent:*

(i) *The space (X,d) is CAT(0).*

(ii) *For any $\mu \in \mathcal{P}^2(X)$, or equivalently for μ discrete, there exists a point $z_\mu \in X$ such that for each $z \in X$ we have*

$$d\left(z,z_\mu\right)^2 + \int_X d\left(x,z_\mu\right)^2 \mathrm{d}\mu(x) \le \int_X d\left(x,z\right)^2 \mathrm{d}\mu(x)\,. \tag{2.3.15}$$

(iii) *We have*

$$\operatorname{var}\mu \le \frac{1}{2}\int_X\int_X d(x,y)^2 \,\mathrm{d}\mu(x)\,\mathrm{d}\mu(y),$$

for every $\mu \in \mathcal{P}^2(X)$, or equivalently for every discrete $\mu \in \mathcal{P}^2(X)$.

(iv) (X, d) *is a length space and for every* $x_1, x_2, x_3, x_4 \in X$ *and* $s, t \in [0, 1]$, *we have*

$$s(s-1)d(x_1, x_3)^2 + t(1-t)d(x_2, x_4)^2 \le std(x_1, x_2)^2 + (1-s)td(x_2, x_3)^2 + (1-s)(1-t)d(x_3, x_4)^2 + s(1-t)d(x_4, x_1)^2 .$$

Proof. Step 1: The variance inequality (2.3.14) immediately gives (i) $\Longrightarrow$ (ii). To show the converse implication (ii) $\Longrightarrow$ (i), choose $x_0, x_1 \in X$ and set $\mu := \frac{1}{2}(\delta_{x_0} + \delta_{x_1})$. Then there exists $z_\mu \in X$ such that for any $z \in X$ we have

$$\frac{1}{2}d(z, x_0)^2 + \frac{1}{2}d(z, x_1)^2 \ge \frac{1}{2}d(z_\mu, x_0)^2 + \frac{1}{2}d(z_\mu, x_1)^2 + d(z, z_\mu)^2 \ge \frac{1}{4}d(x_0, x_1)^2 + d(z, z_\mu)^2 ,$$

where the second inequality follows by (1.3.27). Employing Theorem 1.3.2 gives (i).

Step 2: We prove (ii) $\Longrightarrow$ (iii). By integrating the inequality in (ii) against $d\mu(z)$, we obtain (iii).

We get to (iii) $\Longrightarrow$ (iv). Let us first show that (X, d) is a length space. Choose $x, y \in X$ and $\varepsilon > 0$. Set $\mu := \frac{1}{2}(\delta_x + \delta_y)$. Then (iii) yields a point $m \in X$ such that

$$d(m, x)^2 + d(m, y)^2 \le \frac{1}{2}d(x, y)^2 + \varepsilon.$$

This via Proposition 1.1.2 gives that (X, d) is a length space. To prove the inequality in (iv), set

$$\mu := \frac{1}{2}\left[s\delta_{x_1} + t\delta_{x_2} + (1-s)\delta_{x_3} + (1-t)\delta_{x_4}\right].$$

Then given $\varepsilon > 0$, we know by (iii) that there exists $y \in X$ such that

$$\begin{aligned} &\frac{1}{4}[std(x_1, x_2)^2 + (1-s)td(x_2, x_3)^2 + (1-s)(1-t)d(x_3, x_4)^2 \\ &\quad + s(1-t)d(x_4, x_1)^2 + s(1-s)d(x_1, x_3)^2 + t(1-t)d(x_2, x_4)^2] + \varepsilon \\ &\ge \frac{1}{2}\left[sd(y, x_1)^2 + td(y, x_2)^2 + (1-s)d(y, x_3)^2 + (1-t)d(y, x_4)^2\right] \\ &\ge \frac{1}{2}\left[s(1-s)d(x_1, x_3)^2 + t(1-t)d(x_2, x_4)^2\right] , \end{aligned}$$

where the last inequality follows by (1.3.28). We hence obtained (iv).

Step 3: It remains to show that (iv) $\Longrightarrow$ (i). Given $x, y \in X$ and $s \in (0, 1)$, Proposition 1.1.2 yields a point $m \in X$ such that

$$d(x, m)^2 + d(y, m)^2 \le \frac{1}{2}d(x, y)^2 + s^2.$$

Let $z \in X$ and apply (iv) to $x_1 = z, x_2 = y, x_3 = m, x_4 = x$, and $t = \frac{1}{2}$. Then

$$\begin{aligned} &s(1-s)d(m, z)^2 \\ &\le \frac{s}{2}d(y, z)^2 + \frac{s}{2}d(x, z)^2 + \frac{1-s}{2}d(m, y)^2 + \frac{1-s}{2}d(m, x)^2 - \frac{1}{4}d(x, y)^2 \\ &\le \frac{s}{2}d(y, z)^2 + \frac{s}{2}d(x, z)^2 - \frac{s}{4}d(x, y)^2 + \frac{1-s}{2}s^2 . \end{aligned}$$

Dividing by s and letting $s \to 0$ finishes the proof. □

Lemma 2.3.6. *Let $(\mathcal{H}_i, d_i)_{i\in I}$ be a family of (pointed) Hadamard spaces with an arbitrary index set I. Let $w_i > 0$ for each $i \in I$, and let $(\mathcal{H}, d)$ be the weighted product of $(\mathcal{H}_i, d_i)$ defined in (1.2.10). If $\mu \in \mathcal{P}^2(\mathcal{H})$, then $b(\mu) = (b(\mu_i))_{i\in I}$, where $\mu_i \in \mathcal{P}^2(\mathcal{H}_i)$ are the marginal measures of μ.*

Proof. Put $y := (b(\mu_i))_{i\in I}$. By the definition, $b(\mu)$ is a minimizer of the function $F_y(z) := \int_{\mathcal{H}} [d(z,x)^2 - d(y,x)^2] \, \mathrm{d}\mu(x)$, where $z \in \mathcal{H}$. We have

$$F_y(z) = \int_{\mathcal{H}} \sum_{i\in I} w_i \left[d_i\left(z_i, x_i\right)^2 - d_i((b\left(\mu_i\right), x_i)^2\right] \mathrm{d}\mu(x)$$

$$= \sum_{i\in I} w_i \int_{\mathcal{H}_i} \left[d_i\left(z_i, x_i\right)^2 - d_i((b\left(\mu_i\right), x_i)^2\right] \mathrm{d}\mu\left(x_i\right)$$

and by the variance inequality (2.3.14) we obtain

$$\geq \sum_{i\in I} w_i d_i\left(b\left(\mu_i\right), z_i\right)^2 = d(y,z)^2 ,$$

for any $z \in \mathcal{H}$. It follows that $b(\mu) = y$. □

Lemma 2.3.7. *Let $f : \mathcal{H} \to (-\infty, \infty]$ be a convex lsc function and $\mu \in \mathcal{P}^1(\mathcal{H})$. Then $\int_{\mathcal{H}} f \, \mathrm{d}\mu$ is well defined and is equal to a real number or ∞.*

Proof. Denote $f^+ := \max\{f, 0\}$ and $f^- := \max\{-f, 0\}$. The integral $\int_{\mathcal{H}} f \, \mathrm{d}\mu$ is well defined if $\int f^+ < \infty$ or $\int f^- < \infty$. We claim that one always has $\int f^- < \infty$. Indeed, choose a point $x_0 \in \mathcal{H}$. Lemma 2.2.13 gives $\alpha, \beta \in \mathbb{R}$ such that

$$f(x) \geq \alpha + \beta d\left(x, x_0\right),$$

for every $x \in \mathcal{H}$. Without loss of generality assume $\alpha = 0$ and $\beta < 0$. Then $f^-(x) \leq -\beta d(x, x_0)$ for every $x \in \mathcal{H}$ and therefore

$$\int_{\mathcal{H}} f^-(x) \, \mathrm{d}\mu(x) \leq \int_{\mathcal{H}} -\beta d\left(x, x_0\right) \mathrm{d}\mu(x) < \infty ,$$

since $\mu \in \mathcal{P}^1(\mathcal{H})$. □

Theorem 2.3.8 (Jensen inequality). *Let $(\mathcal{H}, d)$ be a Hadamard space and $\mu \in \mathcal{P}^1(\mathcal{H})$. For a convex lsc function $f : \mathcal{H} \to (-\infty, \infty]$ we have*

$$f\left(b(\mu)\right) \leq \int_{\mathcal{H}} f(x) \, \mathrm{d}\mu(x) . \tag{2.3.16}$$

Proof. Put $\widehat{f}(x) := (x, f(x))$ for every $x \in \mathcal{H}$, and let $\widehat{\mu}$ be the image of μ under $\widehat{f}$.

Step 1: Let us first assume that f is bounded from below. Without loss of generality we can then assume $\int_{\mathcal{H}} |f(x)|\,\mathrm{d}\mu(x) < \infty$, for otherwise the right hand side of (2.3.16) is infinite. Then $\widehat{\mu} \in \mathcal{P}^1(\mathcal{H} \times \mathbb{R})$. Indeed, for any $\widehat{z} := (z, t) \in \mathcal{H} \times \mathbb{R}$ we have

$$\begin{aligned}
\int_{\mathcal{H}\times\mathbb{R}} d\left(\widehat{z}, \widehat{x}\right) \mathrm{d}\widehat{\mu}\left(\widehat{x}\right) &= \int_{\mathcal{H}} d\left(\widehat{z}, (x, f(x))\right) \mathrm{d}\mu(x) \\
&= \int_{\mathcal{H}} \left[d(z,x)^2 + |t - f(x)|^2\right]^{\frac{1}{2}} \mathrm{d}\mu(x) \\
&\leq \int_{\mathcal{H}} \left[d(z,x) + |t - f(x)|\right] \mathrm{d}\mu(x) < \infty .
\end{aligned}$$

Applying Lemma 2.3.6 gives

$$b\left(\widehat{\mu}\right) = \left(b(\mu), \int_{\mathcal{H}} f(x)\,\mathrm{d}\mu(x) \right).$$

Since $\operatorname{supp} \widehat{\mu} \subset \operatorname{epi} f$, we have by Lemma 2.3.3 that $b(\widehat{\mu}) \in \operatorname{epi} f$. That is,

$$f\left(b(\mu)\right) \leq \int_{\mathcal{H}} f(x)\,\mathrm{d}\mu(x).$$

Step 2: If f is not bounded from below, define the functions $g_n := \max(f, -n)$, for $n \in \mathbb{N}$, which are convex lsc and bounded from below. By Step 1 and the monotonicity of the approximation,

$$f\left(b(\mu)\right) \leq \lim_{n\to\infty} \int_{\mathcal{H}} g_n(x)\,\mathrm{d}\mu(x) = \int_{\mathcal{H}} f(x)\,\mathrm{d}\mu(x).$$

The proof is complete. □

A stochastic proof of the Jensen inequality will appear in Theorem 7.2.2.

The Jensen inequality in turn implies convexity. Indeed, let $f : \mathcal{H} \to (-\infty, \infty]$ be a measurable function and $x : [0,1] \to \mathcal{H}$ be a geodesic. When we choose $t \in (0,1)$, the measure $\mu := (1-t)\delta_{x_0} + t\delta_{x_1}$ has the barycenter $b(\mu) = x_t$, and the Jensen inequality (2.3.16) gives

$$f\left(x_t\right) \leq (1-t) f\left(x_0\right) + t f\left(x_1\right),$$

which is of course the convexity condition.

We finish this section making a connection with the Wasserstein distance. For the details, the interested reader is referred to [8]. Given $\mu, \nu \in \mathcal{P}(\mathcal{H})$, we say that a measure $\zeta \in \mathcal{P}(\mathcal{H} \times \mathcal{H})$ is a *coupling* of μ and ν if

$$\zeta(A \times \mathcal{H}) = \mu(A), \quad \text{and} \quad \zeta(\mathcal{H} \times A) = \nu(A),$$

for every Borel $A \subset \mathcal{H}$. One such a coupling ζ is the product measure $\mu \otimes \nu$.

Definition 2.3.9 (Wasserstein distance). The *Wasserstein distance* of measures $\mu, \nu \in \mathcal{P}^1(\mathcal{H})$ is given as

$$d^W(\mu,\nu) := \inf \left\{ \int_{\mathcal{H}} \int_{\mathcal{H}} d(x,y)\, \mathrm{d}\zeta(x,y) \colon\ \zeta \text{ is a coupling of } \mu,\nu \right\}.$$

It is well known that $(\mathcal{P}^1(\mathcal{H}), d^W)$ is a complete metric space.

We have the following estimate on barycenters in terms of the Wasserstein distance.

Proposition 2.3.10. *If $\mu, \nu \in \mathcal{P}^1(\mathcal{H})$, then $d(b(\mu), b(\nu)) \le d^W(\mu,\nu)$.*

Proof. Let $\zeta \in \mathcal{P}^1(\mathcal{H} \times \mathcal{H})$ be a coupling of μ and ν. Then $b(\zeta) = (b(\mu), b(\nu))$ by Lemma 2.3.6. Applying Theorem 2.3.8 on the Hadamard space $\mathcal{H} \times \mathcal{H}$ to the convex function $d \colon \mathcal{H} \times \mathcal{H} \to \mathbb{R}$ yields

$$d\left(b(\mu), b(\nu)\right) = d\left(b(\zeta)\right) \le \int_{\mathcal{H}\times\mathcal{H}} d(w)\, \mathrm{d}\zeta(w).$$

Therefore $d(b(\mu), b(\nu)) \le d^W(\mu,\nu)$. □

Let ψ be an isometry between two Hadamard spaces $(\mathcal{H}, d)$ and $(\mathcal{H}', d')$. If $\mu \in \mathcal{P}^1(\mathcal{H})$, then the push-forward $\psi_*(\mu)$ belongs to $\mathcal{P}^1(\mathcal{H}')$ and

$$\psi_* \colon \mathcal{P}^1(\mathcal{H}) \to \mathcal{P}^1\left(\mathcal{H}'\right)$$

is an isometry between the spaces of probability measures equipped with the Wasserstein distances.

The last result is just a special case of Proposition 2.3.10, we record it explicitly for its significance and give an elementary proof. We use Notation 2.1.9.

Corollary 2.3.11. *Given $(w_1, \dots, w_N) \in \Delta_{N-1}$, the mean of points $a_1, \dots, a_N \in \mathcal{H}$ was in Example 2.2.5 defined as*

$$y := \operatorname*{arg\,min}_{x\in\mathcal{H}} \sum_{n=1}^N w_n d\left(x, a_n\right)^2 .$$

Let now $a_1', \dots, a_N' \in \mathcal{H}$ and denote

$$y' := \operatorname*{arg\,min}_{x\in\mathcal{H}} \sum_{n=1}^N w_n d\left(x, a_n'\right)^2 .$$

Then we have

$$d\left(y, y'\right) \le \sum_{n=1}^N w_n d\left(a_n, a_n'\right).$$

Proof. Inequality (1.3.15) yields

$$d\left(a_n, y'\right)^2 + d\left(a_n', y\right)^2 \le d\left(a_n, y\right)^2 + d\left(a_n', y'\right)^2 + 2d\left(y, y'\right) d\left(a_n, a_n'\right),$$

multiplying by w_n and summing up over n from 1 to N further gives

$$\sum_{n=1}^{N} w_n \left[d\left(a_n, y'\right)^2 + d\left(a_n', y\right)^2\right] \le \sum_{n=1}^{N} w_n \left[d\left(a_n, y\right)^2 + d\left(a_n', y'\right)^2\right] + 2d\left(y, y'\right) \sum_{n=1}^{N} w_n d\left(a_n, a_n'\right).$$

By (2.2.3) we have

$$\sum_{n=1}^{N} w_n \left[d\left(a_n, y'\right)^2 + d\left(a_n', y\right)^2\right] \ge \sum_{n=1}^{N} w_n \left[d\left(a_n, y\right)^2 + d\left(a_n', y'\right)^2\right] + 2d\left(y, y'\right)^2.$$

Altogether we obtain

$$d\left(y, y'\right) \le \sum_{n=1}^{N} w_n d\left(a_n, a_n'\right),$$

which finishes the proof. □

Exercises

Exercise 2.1. Find a geodesic metric space (X, d) and a ball $B(x, r) \subset X$ which is not convex.

Exercise 2.2. Let $(\mathcal{H}_1, d_1)$ and $(\mathcal{H}_2, d_2)$ be two Hadamard spaces and $C \subset \mathcal{H}_1 \times \mathcal{H}_2$ be convex. Denote $\pi_2 \colon \mathcal{H}_1 \times \mathcal{H}_2 \to \mathcal{H}_2$ the projection onto the second factor. Show that the set $\pi_2(C)$ is convex.

Exercise 2.3. Prove Lemma 2.2.1.

Exercise 2.4. Any continuous convex function on a Hilbert space is locally Lipschitz. Is it also true in Hadamard spaces?

Exercise 2.5. Show that a Busemann function from Example 2.2.10 is well defined, that is, the limit exists. Furthermore, show it is convex and Lipschitz. Finally, choose a geodesic ray in $\mathbb{R}^2$ and find the corresponding Busemann function and horoballs.

Exercise 2.6. Assume $f \colon \mathbb{R}^n \to \mathbb{R}$ is a twice differentiable function and $\kappa > 0$. Show that the following are equivalent:

(i) The function f is strongly convex with parameter κ.
(ii) Given $p \in \mathbb{R}^n$, each eigenvalue λ of the Hessian matrix $\nabla^2 f(p)$ satisfies $\lambda \geq \kappa$.
(iii) If $x\colon [0,\infty) \to \mathbb{R}^n$ and $y\colon [0,\infty) \to \mathbb{R}^n$ satisfy

$$\dot{x}(t) = -\nabla f(x(t)), \quad \dot{y}(t) = -\nabla f(y(t)), \quad t \in (0,\infty),$$

along with the initial conditions $x(0) = x_0$ and $y(0) = y_0$ for given $x_0, y_0 \in \mathbb{R}^n$, then we have

$$d(x(t), y(t)) \leq \exp(-t\kappa)\, d(x(0), y(0)),$$

for every $t \in [0,\infty)$.

Exercise 2.7. Prove Lemma 2.2.15.

Exercise 2.8. Let $(\mathcal{H}, d)$ be a Hadamard space and $f\colon \mathcal{H} \to (-\infty,\infty]$ be a convex lsc function. Show that the Moreau–Yosida envelope f_λ is convex, but not necessarily lsc.
Hint. Observe that the function $g\colon \mathcal{H} \times \mathcal{H} \to (-\infty,\infty]$ given by

$$g(x,y) := f(x) + \frac{1}{2\lambda} d(x,y)^2$$

is jointly convex because of (1.2.4) and epi $f_\lambda = \pi_2(\text{epi } g)$, where $\pi_2\colon \mathcal{H} \times \mathcal{H} \times \mathbb{R} \to \mathcal{H} \times \mathbb{R}$ is the projection onto the factor $\mathcal{H} \times \mathbb{R}$. Apply Exercise 2.2 with $\mathcal{H}_1 := \mathcal{H}$ and $\mathcal{H}_2 := \mathcal{H} \times \mathbb{R}$. □

Exercise 2.9 (Means in Hilbert spaces). Show, that if H is a Hilbert space, $a_1, \ldots, a_N \in H$ and $(w_1, \ldots, w_N) \in \Delta_{N-1}$, then

$$\operatorname*{arg\,min}_{x \in H} \sum_{n=1}^{N} w_n d(x, a_n)^2 = w_1 a_1 + \cdots + w_N a_N. \tag{2.3.17}$$

In other words, the mean coincides with the arithmetic mean.

Exercise 2.10. Find an alternative proof of the variance inequality (2.3.14) for a measure $\mu \in \mathcal{P}^2(\mathcal{H})$.
Hint. See (7.1.2). □

Exercise 2.11. Let $(\mathcal{H}, d)$ be a Hadamard space. Show that an arbitrary family of bounded closed convex sets has the finite intersection property.
Hint. Finite subsets of the index set form a directed set. Apply Proposition 2.1.16 to the appropriate net of sets. □

Exercise 2.12. Prove that the closure of co S equals to $\overline{\text{co}}\, S$ for any subset S of a Hadamard space.

Bibliographical remarks

Convexity in Hadamard manifolds was systematically studied in [43]. Geometric aspects of convexity in Hadamard spaces appear throughout the whole Bridson–Haefliger book [51]. A more analytical perspective was presented by J. Jost [104].

Lemma 2.1.8 comes from [102, Lemma 2.6]. The formal convex hull was introduced by U. Lang and V. Schroeder in [127] under the name *correspondence*. Theorem 2.1.12 appeared in [51, Proposition 2.4, p.176] and in the Hadamard manifold case in [43, Lemma 3.2] and [74, Proposition 1.6.3]. We follow Sturm's exposition [195, Proposition 2.6]. Firmly nonexpansive mappings (Definition 2.1.13) in the Hilbert ball are defined on page 124 of the Goebel–Reich book [92]. More recently, they were used in the paper [119] by E. Kopecká and S. Reich. Proposition 2.1.15 mimics the linear version from [166, Theorem 5.6]; see also [90, p. 247]. The fact that there exists a nonconvex Chebyshev set in the hyperbolic plane was pointed out by Genaro López–Acedo. Proposition 2.1.16 comes from [88, Lemma 2.2] and the statement in Exercise 2.11 from [126, Lemma 2.2]. The Hilbert ball case of Proposition 2.1.16 appears in [92, Theorem 18.1]. Optimization problems with the objective function as in Example 2.2.6 have been recently studied from various perspectives [16, 145–147]. The estimate in Lemma 2.2.13 is a nonlinear variant of [13, Lemma 1.5]. Lemma 2.2.19 comes from [88, Lemma 2.2] and Proposition 2.2.27 from [7, Theorem 3.9]. See also [92, Theorem 18.1].

The study of resolvents in Hadamard spaces was initiated by J. Jost in [103] and independently by U. Mayer [139]. The fact that resolvents are nonexpansive (Theorem 2.2.22) was showed in [103, Lemma 4] and [139, Lemma 1.12]. Lemma 2.2.23 is well known; see for instance [8, (4.1.2)]. The resolvent identity in Proposition 2.2.24 is taken from [139, Lemma 1.10]. A similar form of the resolvent identity appeared in [105]. Theorem 2.2.25 is due to J. Jost [103, Theorem 1] and was used to obtain the existence of harmonic maps. For a recent application of the Hamilton–Jacobi semigroup in length spaces, see [134]. The resolvent of a convex lsc function on a Hilbert space is a standard tool in convex analysis [31].

Section 2.3 comes from [195]. Since this paper leaves no space for improvement, we follow its results word by word. It is worth mentioning that the definition of a barycenter given therein applies to $\mathcal{P}^1$-measures, not just $\mathcal{P}^2$-measures. J. Jost was the first to define a barycenter in Hadamard spaces [102, Definition 2.3] and [104, Definition 3.2.1]. The proof of Corollary 2.3.11 is from [126, Lemma 4.2].

3 Weak convergence in Hadamard spaces

Let $(\mathcal{H}, d)$ be a Hadamard space. So far we have considered only the topology induced by the metric d, in which a sequence $(x_n) \subset \mathcal{H}$ converges to a point $x \in \mathcal{H}$ if $d(x, x_n) \to 0$. This convergence is occasionally referred to as *strong* and denoted $x_n \to x$. It turns out however that if the space is not locally compact, some sequences which do not converge strongly at least converge in a weaker sense and in many cases this weak convergence is sufficient to obtain a desired result. We will encounter this phenomenon chiefly in Chapters 5 and 6.

The situation is completely analogous to strong and weak convergences in Hilbert spaces. In particular, we will show that

- any bounded sequence has a weakly convergent subsequence,
- any convex closed set is (sequentially) weakly closed, and
- any convex lsc function is (sequentially) weakly lsc.

On the other hand it is not known whether on a given Hadamard space there exists a topology which induces this type of convergence; see Question 3.1.8. Note that we will work with *sequences* even though it causes no troubles to replace sequences by *nets*.

3.1 Existence of weak limits

Our first important result is the fact that a bounded sequence has a weakly convergent subsequence. Let $(\mathcal{H}, d)$ be a Hadamard space and $(x_n) \subset \mathcal{H}$ be a bounded sequence. Define the function $\omega\colon \mathcal{H} \to [0, \infty)$ as

$$\omega(x; (x_n)) := \limsup_{n \to \infty} d(x, x_n)^2, \quad x \in \mathcal{H}. \tag{3.1.1}$$

This function is strongly convex (see Example 2.2.8) and therefore, by Proposition 2.2.17, it has a unique minimizer, which is called the *asymptotic center* of (x_n).

Definition 3.1.1 (Weak convergence). We shall say that $(x_n) \subset \mathcal{H}$ *weakly converges* to a point $x \in \mathcal{H}$ if x is the asymptotic center of each subsequence of (x_n). We use the notation $x_n \xrightarrow{w} x$.

Clearly, if $x_n \to x$, then $x_n \xrightarrow{w} x$. If there is a subsequence (x_{n_k}) of (x_n) such that $x_{n_k} \xrightarrow{w} z$ for some $z \in \mathcal{H}$, we say that z is a *weak cluster point* of the sequence (x_n).

It is convenient to denote $r(x_n) := \inf_{x \in \mathcal{H}} \omega(x; (x_n))$.

Proposition 3.1.2. *Each bounded sequence has a weakly convergent subsequence, or in other words, each bounded sequence has a weak cluster point.*

Proof. If (u_n) is a subsequence of (v_n), we will use the notation $(u_n) \preccurlyeq (v_n)$. Let $(x_n) \subset \mathcal{H}$ be a bounded sequence. Denote

$$\rho_0 := \inf \{r(v_n) : (v_n) \preccurlyeq (x_n)\},$$

and select $(v_n^1) \preccurlyeq (x_n)$ such that

$$r\left(v_n^1\right) < \rho_0 + 1.$$

Denote

$$\rho_1 := \inf \left\{r(v_n) : (v_n) \preccurlyeq \left(v_n^1\right)\right\}.$$

Having $(v_n^i) \preccurlyeq (v_n^{i-1})$, set

$$\rho_i := \inf \left\{r(v_n) : (v_n) \preccurlyeq \left(v_n^i\right)\right\}.$$

Select $(v_n^{i+1}) \preccurlyeq (v_n^i)$ such that

$$r\left(v_n^{i+1}\right) \leq \rho_i + \frac{1}{i+1}.$$

Since (ρ_n) is nondecreasing and bounded from above by $r(x_n)$, it has a limit, say ρ.

Now take the diagonal sequence (v_k^k) and fix $i \in \mathbb{N}$. Then (v_k^k) is a subsequence (modulo the first $i-1$ elements) of (v_n^i), and hence $r(v_k^k) \geq \rho_i$. On the other hand, for the same fixed $i \in \mathbb{N}$, we have that (v_k^k) is a subsequence (modulo the first i elements) of (v_n^{i+1}), which gives $r(v_k^k) \leq \rho_i + \frac{1}{i+1}$. Taking the limit $i \to \infty$ gives $r(v_k^k) = \rho$.

Since any subsequence (u_n) of (v_k^k) also (for the same reasons) satisfies the inequalities $r(v_k^k) \geq \rho_i$, and $r(v_k^k) \leq \rho_i + \frac{1}{i+1}$, for any $i \in \mathbb{N}$, one gets

$$r(u_n) = \rho\,. \tag{3.1.2}$$

One can conclude that (v_k^k) is the desired subsequence. We also know that there exists a unique point $x \in X$ such that $\limsup_{k\to\infty} d(x, v_k^k)^2 = \rho$. By (3.1.2) we get $v_k^k \xrightarrow{w} x$. □

Proposition 3.1.3. *For a bounded sequence $(x_n) \subset \mathcal{H}$ and $x \in \mathcal{H}$, the following are equivalent:*

(i) *The sequence (x_n) weakly converges to x.*
(ii) *For every geodesic $\gamma\colon [0,1] \to \mathcal{H}$ with $x \in \gamma$, we have $P_\gamma x_n \to x$ as $n \to \infty$.*
(iii) *For every $y \in \mathcal{H}$, we have $P_{[x,y]} x_n \to x$ as $n \to \infty$.*

Proof. (i) $\Longrightarrow$ (ii): Let $\gamma\colon [0,1] \to \mathcal{H}$ be a geodesic with $x \in \gamma$. If

$$\limsup_{n\to\infty} d\left(x, P_\gamma x_n\right) > 0,$$

then there exists a subsequence (y_n) of (x_n) such that $P_\gamma y_n$ converges to some $y \in \gamma \setminus \{x\}$. But then

$$\limsup_{n\to\infty} d\left(y, y_n\right)^2 = \limsup_{n\to\infty} d\left(P_\gamma y_n, y_n\right)^2 \leq \limsup_{n\to\infty} d\left(x, y_n\right)^2,$$

as a consequence of (2.1.1). It contradicts $y_n \xrightarrow{w} x$.

(ii) $\Longrightarrow$ (iii): Trivial.

(iii) $\Longrightarrow$ (i): If the sequence (x_n) does not converge weakly to x, then there exists a subsequence (y_n) of (x_n) such that

$$\limsup_{n\to\infty} d(y, y_n)^2 < \limsup_{n\to\infty} d(x, y_n)^2,$$

for some $y \in \mathcal{H} \setminus \{x\}$. But then

$$\limsup_{n\to\infty} d\left(x, P_{[x,y]} y_n\right) > 0,$$

which contradicts (iii). □

One can easily see from Proposition 3.1.3 that in Hilbert spaces, the notion of weak convergence defined above coincides with the classical weak convergence; see Exercise 3.1. In the following series of lemmas we extend various properties of the weak convergence from Hilbert spaces into Hadamard spaces.

Lemma 3.1.4 (Opial property). *Let $(x_n) \subset \mathcal{H}$ be a sequence weakly converging to a point $x \in \mathcal{H}$. Then we have*

$$\liminf_{n\to\infty} d(x_n, x) < \liminf_{n\to\infty} d(x_n, z)$$

for every $z \in \mathcal{H} \setminus \{x\}$.

Proof. Follows from Proposition 3.1.3. □

Theorem 3.1.5 is an analog of the weak Banach–Saks property in Hilbert spaces, which uses means of points defined in Example 2.2.5 to replace Cesàro means in Hilbert space. The proof relies upon the Opial property from Lemma 3.1.4.

Theorem 3.1.5 (Weak Banach–Saks property). *Let $(x_n) \subset \mathcal{H}$ be a sequence weakly converging to a point x. Then it has a subsequence (x_{n_k}) such that the sequence of the barycenters*

$$b\left(\frac{1}{k}\sum_{i=1}^{k} \delta_{x_{n_i}}\right), \quad k \in \mathbb{N}$$

converges to x as $k \to \infty$.

Proof. Select a subsequence of (x_n), still denoted (x_n), such that $d(x, x_n) \to r$, for some $r \in [0, \infty)$. If $r = 0$, we are done. Let us hence assume $r > 0$. Also observe that

$$\lim_{n\to\infty} \frac{1}{n} \sum_{i=1}^{n} d\left(x, x_i\right)^2 = r^2 .$$

We will proceed in two steps.

Step 1: Given a finite set $I \subset \mathbb{N}$, we denote

$$V(I) := \operatorname{var}\left(\frac{1}{|I|}\sum_{i\in I}\delta_{x_i}\right),$$

where $|I|$ stands for the cardinality of I. Recall that the variance of a measure was defined in Definition 2.3.4. Put $I_k^N := \{(k-1)2^N, \ldots, k2^N\} \subset \mathbb{N}$, for every $k, N \in \mathbb{N}$. We claim that if

$$\sup_{N\in\mathbb{N}} \liminf_{k\to\infty} V\left(I_k^N\right) = r^2, \tag{3.1.3}$$

then

$$b_n := b\left(\frac{1}{n}\sum_{i=1}^{n}\delta_{x_i}\right), \quad n \in \mathbb{N}$$

converges to x. Indeed, for every $\varepsilon > 0$, there exists $N \in \mathbb{N}$ such that

$$\liminf_{k\to\infty} V\left(I_k^N\right) \geq r^2 - \varepsilon.$$

Then we have

$$\begin{aligned}
r^2 &\geq \limsup_{n\to\infty} \frac{1}{n}\sum_{i=1}^{n} d\left(b_n, x_i\right)^2 \\
&\geq \liminf_{n\to\infty} \frac{1}{n}\sum_{i=1}^{n} d\left(b_n, x_i\right)^2 \\
&\geq \liminf_{k\to\infty} \frac{1}{k}\sum_{i=1}^{k} V\left(I_k^N\right) \\
&\geq r^2 - \varepsilon.
\end{aligned}$$

Since ε was arbitrary, we get $\lim_{n\to\infty} \frac{1}{n}\sum_{i=1}^{n} d(b_n, x_i)^2 = r^2$. The variance inequality (2.3.14) reads

$$\frac{1}{n}\sum_{i=1}^{n} d\left(x, x_i\right)^2 - \frac{1}{n}\sum_{i=1}^{n} d\left(b_n, x_i\right)^2 \geq d\left(b_n, x\right)^2,$$

and immediately implies $b_n \to x$.

Step 2: We will select a subsequence of (x_n) which ultimately satisfies assumption (3.1.3) in the previous step. Put $J_k^0 := \{k\}$ for each $k \in \mathbb{N}$, and construct inductively, for every $N \in \mathbb{N}$, a sequence $(J_k^N)_{k\in\mathbb{N}}$ of subsets of $\mathbb{N}$ of cardinality 2^N such that $J_k^N = J_l^{N-1} \cup J_m^{N-1}$ for some $l, m \in \mathbb{N}$, and $\max J_k^N < \min J_{k+1}^N$, and $J_1^N \subset J_1^{N+1}$, and

$$\lim_{k\to\infty} V\left(J_k^N\right) = V_N := \limsup_{l,m\to\infty} V\left(J_l^{N-1} \cup J_m^{N-1}\right).$$

It is easy to see that $V_N \leq V_{N+1} \leq r^2$ for every $N \in \mathbb{N}$. We claim that $V_N \to r^2$ as $N \to \infty$. To prove this, we will show that for each $\varepsilon \in (0, r)$ there exists $\delta > 0$ such that if $V_N < (r-\varepsilon)^2$ for some $N \in \mathbb{N}$, then $V_{N+1} > V_N + \delta$. Fix $N \in \mathbb{N}$ and for each $l \in \mathbb{N}$, denote by b_l^N the barycenter of the measure

$$\frac{1}{2^N} \sum_{i \in J_l^N} \delta_{x_i} .$$

Fix $l \in \mathbb{N}$ such that $V(J_i^N) < (r-\varepsilon)^2$ for each $i > l$. Then choose $m > l$ such that $d(b_l^N, x_i) > r$ for every $i \in J_m^N$ on account of Lemma 3.1.4. We arrive at

$$\frac{1}{2^N} \sum_{i \in J_m^N} d\left(b_l^N, x_i\right)^2 > r^2 > (r-\varepsilon)^2 > V\left(J_m^N\right) = \frac{1}{2^N} \sum_{i \in J_m^N} d\left(b_m^N, x_i\right)^2 ,$$

and consequently,

$$2 \max\left\{d\left(b_m^N, b_{l\cup m}^{N+1}\right), d\left(b_{l\cup m}^{N+1}, b_l^N\right)\right\} \geq d\left(b_m^N, b_l^N\right) > \varepsilon ,$$

where $b_{l\cup m}^{N+1}$ stands for the barycenter of the measure

$$\frac{1}{2^{N+1}} \sum_{i \in J_l^N \cup J_m^N} \delta_{x_i} .$$

Invoking the variance inequality (2.3.14) again, we obtain

$$\begin{aligned} V\left(J_l^N \cup J_m^N\right) &= \frac{1}{2^{N+1}} \left[\sum_{i \in J_l^N} d\left(b_{l\cup m}^{N+1}, x_i\right)^2 + \sum_{i \in J_l^N} d\left(b_{l\cup m}^{N+1}, x_i\right)^2 \right] \\ &\geq \frac{1}{2} \left[V\left(J_l^N\right) + V\left(J_m^N\right) + \frac{\varepsilon^2}{4} \right] . \end{aligned}$$

We have just proved $V_N \to r^2$ as $N \to \infty$. Finally, observe that the set $\bigcap_N \bigcup_k J_k^N \subset \mathbb{N}$ is infinite. Let us denote its elements in increasing order by $(n_1, n_2, \ldots)$. Then the sequence (x_{n_k}) satisfies assumption (3.1.3) in Step 1, modulo renaming (x_{n_k}) to (x_n).

The proof is hence complete. □

The conclusion of Proposition 3.1.6 is a counterpart of the *Kadec–Klee property* in Banach spaces. Recall that a Banach space has the Kadec–Klee property if the weak and strong convergences of sequences coincide on the unit sphere of the Banach space [82]. Hilbert spaces of course enjoy this property.

Proposition 3.1.6 (Kadec–Klee property). *Let $(x_n) \subset \mathcal{H}$ and $x \in \mathcal{H}$. Then $x_n \to x$ if and only if $x_n \xrightarrow{w} x$ and $d(x_n, y) \to d(x, y)$ for some $y \in \mathcal{H}$.*

Proof. Let $x_n \xrightarrow{w} x$ and $d(x_n, y) \to d(x, y)$ for some $y \in \mathcal{H}$. Note that $d(x_n, y)^2 \geq d(x_n, P_{[x,y]}x_n)^2 + d(P_{[x,y]}x_n, y)^2$ holds. Since $P_{[x,y]}x_n$ converges to x by Proposition 3.1.3, we have $d(x_n, P_{[x,y]}x_n)^2 \to 0$ and consequently $x_n \to x$. The converse implication is trivial. □

A somewhat quantified version of Proposition 3.1.6 is in the following result.

Proposition 3.1.7 (Uniform Kadec–Klee property). *For every $r > 0$ and $\varepsilon \in (0, r)$ there exists $\delta \in (0, r)$ such that for every $y \in \mathcal{H}$ and a sequence $(x_n) \subset B(y, r)$ weakly converging to a point $x \in \mathcal{H}$ and satisfying*

$$\limsup_{n\to\infty} d(x, x_n) > \varepsilon ,$$

we have $d(x, y) \leq r - \delta$.

Proof. Select a subsequence of (x_n), still denoted (x_n), such that $d(x, x_n) > \varepsilon$ for each $n \in \mathbb{N}$. Given $\lambda > 0$, there exists $n_0 \in \mathbb{N}$ so that for every $n > n_0$, we have

$$d(y, x_n)^2 \geq d(x, y)^2 + d(x, x_n)^2 - \lambda ,$$

on account of (2.1.1). Then $d(x, y)^2 \leq r^2 - \varepsilon^2 + \lambda$, and since λ was arbitrary, we have $d(x, y)^2 \leq r^2 - \varepsilon^2$. □

The existence of weak convergence raises the following question.

Question 3.1.8 (Weak topology). Is there a topology τ on $(\mathcal{H}, d)$ such that for each bounded sequence $(x_n) \subset \mathcal{H}$ and a point $x \in \mathcal{H}$ we have $x_n \xrightarrow{w} x$ if and only if $x_n \xrightarrow{\tau} x$?

In Chapters 4 and 5 we will need the notion of the weak convergence of a *curve* to a point.

Definition 3.1.9 (Weak convergence of a curve). Let $(\mathcal{H}, d)$ be a Hadamard space and $c\colon [0, \infty) \to \mathcal{H}$ be a curve (that is, a continuous mapping). Then c is said to *weakly converge* to a point $x \in \mathcal{H}$ if $c(t_n) \xrightarrow{w} x$ for each sequence $(t_n) \subset [0, \infty)$ with $t_n \to \infty$. We then write $c(t) \xrightarrow{w} x$.

Note that the previous definition implicitly presumes that c is bounded. Alternatively, one might view a curve as a net.

3.2 Weak convergence and convexity

We now get to the remaining two properties of weak convergence promised in the beginning of the present chapter. Let again $(\mathcal{H}, d)$ be a Hadamard space.

Lemma 3.2.1. *Let $C \subset \mathcal{H}$ be a closed convex set and $(x_n) \subset C$. If the sequence (x_n) weakly converges to a point $x \in \mathcal{H}$, then $x \in C$.*

Proof. Assume that $x \notin C$ and denote $\gamma := [x, P_C x]$. We claim that $P_\gamma x_n = P_C x$ for each $n \in \mathbb{N}$. Indeed, if for some $m \in \mathbb{N}$ we had $P_\gamma x_m \neq P_C x$, then by (2.1.1), we would have both

$$d\left(x_m, P_\gamma x_m\right)^2 \geq d\left(x_m, P_C x\right)^2 + d\left(P_C x, P_\gamma x_m\right)^2,$$

and

$$d\left(x_m, P_C x\right)^2 \geq d\left(x_m, P_\gamma x_m\right)^2 + d\left(P_C x, P_\gamma x_m\right)^2,$$

which is impossible. Finally,

$$d(P_\gamma x_n, x) = d\left(P_C x, x\right) \neq 0,$$

for each $n \in \mathbb{N}$, which, by Proposition 3.1.3, contradicts $x_n \xrightarrow{w} x$. □

Definition 3.2.2 (Weak lower semicontinuity). We shall say that a function $f \colon \mathcal{H} \to (-\infty, \infty]$ is *weakly lsc* at a given point $x \in \operatorname{dom} f$ if

$$\liminf_{n\to\infty} f(x_n) \geq f(x),$$

for each sequence $x_n \xrightarrow{w} x$. We say that f is *weakly lsc* if it is weakly lsc at every $x \in \operatorname{dom} f$.

Lemma 3.2.3. *If $f \colon \mathcal{H} \to (-\infty, \infty]$ is a convex lsc function, then it is weakly lsc.*

Proof. By contradiction. Let $(x_n) \subset \mathcal{H}$ be a sequence weakly converging to a point $x \in \operatorname{dom} f$. Suppose that

$$\liminf_{n\to\infty} f(x_n) < f(x).$$

That is, there exist a subsequence (x_{n_k}), index $k_0 \in \mathbb{N}$ and $\delta > 0$ such that $f(x_{n_k}) < f(x) - \delta$ for each $k > k_0$. By the lower semicontinuity and convexity of f, we get

$$f(y) \leq f(x) - \delta$$

for every $y \in \overline{\operatorname{co}}\{x_{n_k} : k > k_0\}$. But this, through Lemma 3.2.1, yields a contradiction to $x_n \xrightarrow{w} x$. □

Corollary 3.2.4. *Let $C \subset \mathcal{H}$ be a closed convex set. The distance function d_C as well as its square d_C^2 are weakly lsc.*

Next we introduce a simple, but useful notion.

Definition 3.2.5 (Fejér monotone sequence). A sequence $(x_n) \subset \mathcal{H}$ is *Fejér monotone* with respect to a set $S \subset \mathcal{H}$ if, for each $y \in S$, we have

$$d\left(x_{n+1}, y\right) \leq d\left(x_n, y\right), \qquad n \in \mathbb{N}.$$

Recall that a sequence $(x_n) \subset \mathcal{H}$ is said to converge *linearly* to a point $x \in \mathcal{H}$ if there exist $L \geq 0$ and $\theta \in [0, 1)$ such that

$$d\left(x, x_n\right) \leq L\theta^n, \quad n \in \mathbb{N}. \tag{3.2.4}$$

The parameter θ is called the *rate* of linear convergence.

Proposition 3.2.6 (Properties of Fejér monotonicity). *Let $C \subset \mathcal{H}$ be a convex closed set and $(x_n) \subset \mathcal{H}$ be a Fejér monotone sequence with respect to C. Then the following holds:*

(i) *The sequence (x_n) is bounded.*
(ii) $d(x_{n+1}, C) \leq d(x_n, C)$ *for each* $n \in \mathbb{N}$.
(iii) *The sequence (x_n) weakly converges to some $x \in C$ if and only if all weak cluster points of (x_n) belong to C.*
(iv) *The sequence (x_n) converges to some $x \in C$ if and only if $d(x_n, C) \to 0$.*
(v) *The sequence (x_n) converges linearly to some $x \in C$, provided there exists $\theta \in [0, 1)$ such that $d(x_{n+1}, C) \leq \theta d(x_n, C)$ for each $n \in \mathbb{N}$.*

Proof. (i) and (ii) are easy. Let us prove the nontrivial implication of (iii). Assume that all weak cluster points of (x_n) lie in C. It suffices to show that (x_n) has a unique cluster point. By contradiction, let $c_1, c_2 \in C$, with $c_1 \neq c_2$, be weak cluster points of (x_n). That is, there are subsequences (x_{n_k}) and (x_{m_k}) such that $x_{n_k} \xrightarrow{w} c_1$ and $x_{m_k} \xrightarrow{w} c_2$. Without loss of generality, assume $r(x_{n_k}) \leq r(x_{m_k})$. For any $\varepsilon > 0$ there exists $k_0 \in \mathbb{N}$ such that $d(x_{n_k}, c_1)^2 < r(x_{n_k}) + \varepsilon$, for each $k \geq k_0$. By Fejér monotonicity we also have $d(x_{m_k}, c_1)^2 < r(x_{n_k}) + \varepsilon$, for every $m_k \geq n_{k_0}$. Hence, there exists $k_1 \in \mathbb{N}$ such that $d(x_{m_k}, c_1) < r(x_{m_k}) + \varepsilon$, for each $k \geq k_1$. But this contradicts the fact that c_2 is the unique asymptotic center of (x_{m_k}).

Now we prove (iv). Suppose $d(x_n, C) \to 0$. Since for each $k \in \mathbb{N}$ we have

$$d(x_{n+k}, x_n) \leq d\left(x_{n+k}, P_C x_n\right) + d\left(x_n, P_C x_n\right) \tag{3.2.5a}$$

and hence, by Fejér monotonicity,

$$\leq d\left(x_n, P_C x_n\right) + d\left(x_n, P_C x_n\right) \leq 2d(x_n, C), \tag{3.2.5b}$$

we obtain that (x_n) is Cauchy and so converges to a point from C. The converse implication in (iv) is trivial. It remains to prove (v). From (3.2.5) we get

$$d(x_{n+k}, x_n) \leq 2d(x_n, C) \leq 2\theta^n d(x_0, C)$$

for every $n, k \in \mathbb{N}$. The sequence (x_n) obviously converges to some $x \in C$ and thus letting $k \to \infty$ we have

$$d(x, x_n) \leq 2d(x_n, C) \leq 2\theta^n d(x_0, C), \qquad n \in \mathbb{N}.$$

In other words, $x_n \to x$ linearly with rate θ, completing the proof. □

3.3 An application in fixed point theory

As the first application of weak convergence in Hadamard spaces, we present the following fixed point theorem.

Theorem 3.3.1. *Let $(\mathcal{H}, d)$ be a Hadamard space and $C \subset \mathcal{H}$ be convex, closed and bounded. Then for any $F: C \to C$ nonexpansive, the set of fixed points* $\operatorname{Fix} F$ *is convex, closed and nonempty.*

Proof. We start by proving the existence of a fixed point.

Step 1: We first show that $\inf_{x\in C} d(x, Fx) = 0$. Choose $\varepsilon \in (0,1)$ and $z \in C$. Define a mapping $G: C \to C$ by $Gx := (1-\varepsilon)Fx + \varepsilon z$ for every $x \in C$. Since F is nonexpansive, the mapping G is a contraction and hence has a fixed point $y \in \mathcal{H}$ by the Banach contraction principle. An easy estimate gives

$$d(y, Fy) = d(Gy, Fy) = \varepsilon d(z, Fy) \leq \varepsilon \operatorname{diam} C,$$

which yields $\inf_{x\in C} d(x, Fx) = 0$.

Step 2: Choose a sequence $x_n \in C$ such that $d(x_n, Fx_n) \to 0$. Since C is bounded, there exists a subsequence of (x_n), still denoted by (x_n), weakly converging to some $x \in C$. Let us now estimate

$$\begin{aligned}\limsup_{n\to\infty} d\left(Fx, x_n\right) &\leq \limsup_{n\to\infty} \left[d\left(Fx, Fx_n\right) + d\left(Fx_n, x_n\right)\right] \\ &= \limsup_{n\to\infty} d\left(Fx, Fx_n\right) \\ &\leq \limsup_{n\to\infty} d\left(x, x_n\right) .\end{aligned}$$

By the uniqueness of the weak limit we get $x = Fx$.

Step 3: It has already been observed in Example 2.1.5 that the set of fixed points $\operatorname{Fix} F$ of a nonexpansive mapping $F: \mathcal{H} \to \mathcal{H}$ is convex and closed. □

Approximation methods for fixed points will be studied in Section 6.2.

Exercises

Exercise 3.1 (Hilbert space). Let H be a Hilbert space, and $\langle\cdot,\cdot\rangle$ its inner product. A sequence $(x_n) \subset H$ is said to converge weakly to a point $x \in H$ if

$$\langle x_n - x, y\rangle \to 0, \quad n \to \infty,$$

for each $y \in H$. Show that this weak convergence coincides with the one defined in Definition 3.1.1.

Exercise 3.2 (Hilbert ball). Let $(\mathbb{B}, \rho)$ be the Hilbert ball from Example 1.2.13. Assume that (x_n) is a bounded sequence (with respect to ρ) and $x \in \mathbb{B}$. Show that $x_n \xrightarrow{w} x$ if and only if (x_n) converges to x weakly in the usual the Hilbert space sense.

Hint. See [92, Proposition 21.4] if need be. □

Exercise 3.3 (Projection theorem revisited). Use weak convergence to show that, given $C \subset \mathcal{H}$ a closed convex subset of a Hadamard space $(\mathcal{H}, d)$ and $x \in \mathcal{H}$, there exists a point $c \in C$ such that $d_C(x) = d(c, x)$.
Hint. There exists $(c_n) \subset C$ such that $d(c_n, x) \to d_C(x)$. It is bounded, so a subsequence (c_{n_k}) weakly converges to some $c \in \mathcal{H}$. Since C is convex, $c \in C$. Finally,

$$d_C(x) \leq d(x, c) \leq \liminf_{n \to \infty} d\left(x, c_n\right) = d_C(x). \qquad \square$$

Exercise 3.4 (Weak Banach–Saks property). Let (x_n) be a sequence in a Hilbert space which weakly converges to a point x. Show that it has a subsequence (x_{n_k}) such that the sequence of Cesàro means

$$a_k := \frac{x_{n_1} + \cdots + x_{n_k}}{k}, \quad k \in \mathbb{N}$$

converges to x.

Bibliographical remarks

Weak convergence is an essential notion in Banach spaces [82]. J. Jost was the first to introduce weak convergence in Hadamard spaces [102, Definition 2.7]. It was later rediscovered in the following development. E. N. Sosov [188] introduced ψ- and φ-convergences, both generalizing Hilbert space weak convergence into geodesic metric spaces, which motivated T. C. Lim [132] to define his Δ-convergence. It was then brought into Hadamard spaces by W. A. Kirk and B. Panyanak [111] and finally, R. Espínola and A. Fernández–León [81] modified Sosov's φ-convergence to obtain an equivalent formulation of Δ-convergence in Hadamard spaces, which however was exactly the original weak convergence due to J. Jost [102].

Weak convergence theory was in a similar form surveyed in [17, Section 3]. Unlike [111], we use in (3.1.1) the square power of the distance function, which immediately yields a unique minimizer of the function ω. The existence and uniqueness of an asymptotic center in the Hilber ball was proved in [92, Proposition 21.1]. The proof of Proposition 3.1.2 mimics [91, Lemma 15.2]. J. Jost [102, Theorem 2.1] gave a different proof of Proposition 3.1.2. The characterization of weak convergence in Proposition 3.1.3 comes from [81, Proposition 5.2]. The property in Lemma 3.1.4 is named after Z. Opial [156] who established it for Hilbert spaces. The weak Banach–Saks property is known to be enjoyed by Hilbert spaces [82]. Its counterpart in Hadamard spaces was asserted in [102, Theorem 2.2]. In Theorem 3.1.5 we reproduce the proof of [209, Theorem C] due to T. Yokota. Proposition 3.1.7 was inspired by [111, Theorem 3.9] and [17, Proposition 3.6]. Lemmas 3.2.1 and 3.2.3 appeared in [22] and [19], respectively. The Hilbert ball case of Lemma 3.2.1 can be deduced from [92, Lemma 21.2].

Question 3.1.8 was made explicit in [111, p. 3696].

P. Combettes [67] surveys the importance and usefulness of Fejér monotonicity in optimization and also describes its history. In the context of Hadamard spaces, this property was first used in [22]. Proposition 3.2.6 is from [22, Proposition 3.3].

It is also possible to define weak convergence in an asymptotic relation of Hadamard spaces [124, Definition 5.2]. See also [17, Section 3].

Theorem 3.3.1 is due to W. A. Kirk; see [112, Theorem 18] and [110, Theorem 12]. For a nice application in graph theory, see [109].

I am grateful to Takumi Yokota for informing me of his work [209].

4 Nonexpansive mappings

Let $(\mathcal{H}, d)$ be a Hadamard space. A mapping $F\colon \mathcal{H} \to \mathcal{H}$ is *nonexpansive* if $d(Fx, Fy) \leq d(x, y)$ for every $x, y \in \mathcal{H}$. We have already met such mappings in previous chapters, for instance in Theorem 3.3.1. The metric projection in Theorem 2.1.12 and the resolvent of a function in Definition 2.2.20 are important examples of nonexpansive mappings in Hadamard spaces.

In this chapter, we first present an extension theorem in the spirit of the Kirszbraun–Valentine theorem, then introduce a resolvent of a nonexpansive mapping and finally we will show that the celebrated Crandall–Liggett construction of semigroups carries over into Hadamard spaces. The results developed here are the crucial building blocks in the proof of the Lie–Trotter–Kato formula in Section 5.3. We believe though, they are interesting on their own.

4.1 Kirszbraun–Valentine extension

In this section we are concerned with the question as to whether it is possible to generalize the Kirszbraun–Valentine theorem for mappings with values in a Hadamard space. More precisely, given a nonexpansive mapping defined on a subset of a Hilbert space with values in a Hadamard space, does there exist a nonexpansive extension of this mapping defined on the whole Hilbert space?

The answer is yes and before presenting this nice result, we recall the notion of a tangent cone from Section 1.2. Let $(\mathcal{H}, d)$ be a Hadamard space and $p \in \mathcal{H}$. Then the tangent cone $T_p\mathcal{H}$ was defined as the Euclidean cone over the space of directions at p, and the completed tangent cone $\overline{T}_p\mathcal{H}$ is the Euclidean cone over the completed space of directions at p. Also recall that the latter is a Hadamard space when equipped with the cone metric d_p. The point $(0, x) \in \overline{T}_p\mathcal{H}$, with an arbitrary x, is called the origin of $\overline{T}_p\mathcal{H}$ and we denote it by o. Put also $\|v\|_p := d_p(o, v)$ for each $v \in \overline{T}_p\mathcal{H}$, and define an analog of a Riemannian metric on $\overline{T}_p\mathcal{H}$ by setting

$$\langle u, v\rangle_p := \frac{\|u\|^2 + \|v\|^2 - d_p(u, v)^2}{2}, \qquad u, v \in \overline{T}_p\mathcal{H}.$$

One can easily verify that

$$\langle u, v\rangle_p = st \cos \alpha(x, y),$$

whenever $u = (s, x)$ and $v = (t, y)$ and $\alpha(x, y)$ is the angle distance between the directions x and y.

The following lemma will be needed. It uses the notion of a formal convex hull introduced in Definition 2.1.10.

Lemma 4.1.1. *Let $(\mathcal{H}, d)$ be a Hadamard space and $p \in \mathcal{H}$. If $v_1, \dots, v_N \in \overline{T}_p\mathcal{H}$ and $(u, \lambda), (w, \mu) \in \operatorname{fco}\{v_1, \dots, v_N\}$, then*

$$\langle u, w \rangle_p \geq \sum_{i,j=1}^{N} \lambda_i \mu_j \left\langle v_i, v_j \right\rangle_p ,$$

where $\lambda := (\lambda_1, \dots, \lambda_N)$ and $\mu := (\mu_1, \dots, \mu_N)$ are from Δ_{N-1}.

Proof. Step 1: For the moment, let us denote the midpoint of v_1 and v_2 by m. We claim that

$$\langle m, z \rangle_p \geq \frac{1}{2} \langle v_1, z \rangle_p + \frac{1}{2} \langle v_2, z \rangle_p , \tag{4.1.1}$$

for each $z \in \overline{T}_p\mathcal{H}$. Indeed, since we have

$$\|m\|_p^2 = \frac{1}{2} \|v_1\|_p^2 + \frac{1}{2} \|v_2\|_p^2 - d_p(v_1, v_2)^2 ,$$

we arrive at

$$\frac{\|m\|_p^2 + \|z\|_p^2 - d_p(m, z)^2}{2} \geq \frac{\|v_1\|_p^2 + \|z\|_p^2 - d_p(v_1, z)^2}{4} + \frac{\|v_2\|_p^2 + \|z\|_p^2 - d_p(v_2, z)^2}{4}$$

on account of (1.2.3), and the last inequality already implies (4.1.1).

Step 2: Let $\gamma\colon [0, 1] \to \overline{T}_p\mathcal{H}$ be the geodesic from v_1 to v_2. We claim that

$$\langle \gamma_t, z \rangle_p \geq (1 - t) \langle \gamma_0, z \rangle_p + t \langle \gamma_1, z \rangle_p , \tag{4.1.2}$$

for each $z \in \overline{T}_p\mathcal{H}$ and $t \in [0, 1]$. Indeed, we repeatedly use (4.1.1) to prove (4.1.2) for every dyadic $t := k/2^l$, where $k = 1, \dots, 2^l$ and $l \in \mathbb{N}$, and then extend the inequality for an arbitrary $t \in [0, 1]$ by continuity.[1]

Step 3: Fix now $t \in [0, 1]$ and let $\eta\colon [0, 1] \to \overline{T}_p\mathcal{H}$ be the geodesic from γ_t to v_3. Applying (4.1.2) twice, first with γ and then with η, yields

$$\langle \eta_s, z \rangle \geq (1 - t)(1 - s) \langle v_1, z \rangle + t(1 - s) \langle v_2, z \rangle + s \langle v_3, z \rangle ,$$

for each $z \in \overline{T}_p\mathcal{H}$ and $s \in [0, 1]$. Iterating this process gives the proof. □

We are now ready to state the main result of the present section.

Theorem 4.1.2 (Extending nonexpansive mappings). *Let $(H, \|\cdot\|)$ be a Hilbert space and $(\mathcal{H}, d)$ be a Hadamard space. Given a set $S \subset H$ and a nonexpansive mapping $F\colon S \to \mathcal{H}$, there exists a nonexpansive mapping $G\colon H \to \mathcal{H}$ such that $G \restriction_S = F$ and $G(H) \subset \overline{\operatorname{co}}\, F(S)$.*

1 Such an argument was used several times in Chapter 1.

Proof. Step 1: First we show that a nonexpansive mapping defined on a set of N points $x_1, \ldots, x_N \subset H$, where $N \in \mathbb{N}$, can be extended onto the set $\{x_1, \ldots, x_N\} \cup \{x\}$, where $x \in H$. Let hence $F\colon \{x_1, \ldots, x_N\} \to \mathcal{H}$ be nonexpansive and denote $y_n := f(x_n)$ for every $n = 1, \ldots, N$. Since the set of minimizers of the function

$$z \mapsto \max_{n=1,\ldots,N} \frac{d(z, y_n)}{\|x - x_n\|}, \quad z \in \mathcal{H},$$

coincides with the set of minimizers of the function

$$z \mapsto \max_{n=1,\ldots,N} \frac{d(z, y_n)^2}{\|x - x_n\|^2}, \quad z \in \mathcal{H},$$

and the latter is strongly convex, there exists a unique point $y \in \mathcal{H}$ such that

$$y = \arg\min_{z \in \mathcal{H}} \max_{n=1,\ldots,N} \frac{d(z, y_n)}{\|x - x_n\|}. \tag{4.1.3}$$

Here we used Proposition 2.2.17. It is easy to see that $y \in \overline{\operatorname{co}}\{y_1, \ldots, y_N\}$. If we denote

$$L := \max_{n=1,\ldots,N} \frac{d(y, y_n)}{\|x - x_n\|},$$

then obviously $d(y, y_n) \le L\|x - x_n\|$, for each $n = 1, \ldots, N$, and there exist indices for which the equality is attained. Without loss of generality we may assume there exists $1 \le k \le N$ such that $d(y, y_n) = L\|x - x_n\|$, for each $n = 1, \ldots, k$.

Assume that $L > 1$. Let $\gamma_n\colon [0, 1] \to \mathcal{H}$ be the geodesic from y to y_n and let $v_n := (1, \gamma_n) \in T_y\mathcal{H}$, for $n = 1, \ldots, k$. Here we identified the geodesics γ_n's with the directions they determine. We will now show that the origin o of the completed tangent cone $\overline{T}_y\mathcal{H}$ lies in $\overline{\operatorname{co}}\{v_1, \ldots, v_k\}$. If it were not the case, denote by w the projection of o onto $\overline{\operatorname{co}}\{v_1, \ldots, v_k\}$. By Theorems 1.2.17 and 2.1.12 we know that $\alpha(o, w, v_n) \ge \frac{\pi}{2}$ and consequently $\alpha(w, o, v_n) < \frac{\pi}{2}$. Varying w a little, we may assume $w \in T_y\mathcal{H}$, and there therefore exists a geodesic $\eta\colon [0, 1] \to \mathcal{H}$ with $\eta(0) = y$ in the direction of w. Then of course $\alpha(\eta, \gamma_n) = \alpha(w, o, v_n) < \frac{\pi}{2}$, for $n = 1, \ldots, k$. We hence obtain that there exists $t \in (0, 1)$ such that $d(\eta_t, y_n) < L\|x - x_n\|$, for $n = 1, \ldots, k$, which contradicts (4.1.3). We therefore arrive at $o \in \overline{\operatorname{co}}\{v_1, \ldots, v_k\}$.

Given $\varepsilon > 0$, there exists $v' \in \operatorname{co}\{v_1, \ldots, v_k\}$ with $\|v'\|_y \le \varepsilon$. Let $\lambda := (\lambda_1, \ldots, \lambda_k)$ be such that $(v', \lambda) \in \operatorname{fco}\{v_1, \ldots, v_k\}$. By virtue of Lemma 4.1.1 we get

$$\varepsilon^2 \ge \|v'\|_y^2 \ge \sum_{i,j=1}^{k} \lambda_i\lambda_j \langle v_i, v_j\rangle_y = \sum_i \lambda_i^2 + 2\sum_{i<j} \lambda_i\lambda_j \langle v_i, v_j\rangle_y .$$

Since $\sum_i \lambda_i^2 \ge \frac{1}{k}$ and $\langle v_i, v_j\rangle_y \ge -1$, the above inequality yields

$$2\sum_{i<j} \lambda_i\lambda_j \ge \frac{1}{k} - \varepsilon^2 .$$

Next we observe that

$$\left\langle v_i, v_j \right\rangle_y = \cos\alpha\left(y_i, y_j\right) \geq \cos\overline{\alpha}\left(y_i, y, y_j\right),$$

for every $1 \leq i < j \leq k$. Furthermore, since $d(y_i, y_j) \leq \|x_i - x_j\|$ and $d(y, y_i) = L\|x - x_i\| > \|x - x_i\|$, we have $\cos\overline{\alpha}(y_i, y, y_j) > \xi_{ij}$, where

$$\xi_{ij} := \frac{\left\langle x_i - x, x_j - x \right\rangle}{\left\|x_i - x\right\| \left\|x_j - x\right\|},$$

with $\langle\cdot,\cdot\rangle$ standing for the inner product in H. Consequently, there exists $\delta > 0$ such that $\langle v_i, v_j\rangle_y \geq \xi_{ij} + \delta$, for $1 \leq i < j \leq k$. Note that δ is independent of ε. Finally, we arrive at

$$\begin{aligned}
\varepsilon^2 &\geq \sum_i \lambda_i^2 + 2\sum_{i<j} \lambda_i\lambda_j\xi_{ij} + \delta\left(\frac{1}{k} - \varepsilon^2\right) \\
&= \sum_{i,j=1}^{k} \lambda_i\lambda_j\xi_{ij} + \delta\left(\frac{1}{k} - \varepsilon^2\right) \\
&\geq \delta\left(\frac{1}{k} - \varepsilon^2\right),
\end{aligned} \tag{4.1.4}$$

where the last inequality follows by the fact that the Gram matrix $(\xi_{ij})_{i,j=1}^k$ is positive semidefinite. However, if ε is small enough in (4.1.4), we obtain a contradiction. This shows that $L \leq 1$.

Step 2: Let $S \subset H$ and $x \in H$. Given a nonexpansive mapping $F\colon S \to H$, we claim there exists its nonexpansive extension $F'\colon S \cup \{x\} \to \mathcal{H}$. Indeed, it suffices to show

$$\bigcap_{u\in S} B\left(F(u), \|u - x\|\right) \neq \emptyset .$$

At the previous step we proved that

$$\bigcap_{u\in S_0} B\left(F(u), \|u - x\|\right) \neq \emptyset ,$$

for an arbitrary finite set $S_0 \subset S$. Applying Exercise 2.11 then gives the desired claim. It is easy to see that $F'(x) \in \overline{\mathrm{co}}\, F(S)$.

Step 3: Denote by $\mathcal{E}$ the set of nonexpansive extensions $F'\colon S' \to \mathcal{H}$ of F with $S \subset S' \subset H$ and $F'(S') \subset \overline{\mathrm{co}}\, F(S)$. We define an order on the set $\mathcal{E}$ by

$$F_1' \preccurlyeq F_2' \quad \text{if } S_1' \subset S_2' \text{ and } F_2' \restriction_{S_1'} = F_1' ,$$

for every $F_1'\colon S_1' \to \mathcal{H}$ and $F_2'\colon S_2' \to \mathcal{H}$ from $\mathcal{E}$. Since every chain in $\mathcal{E}$ has an upper bound, Zorn's lemma yields a maximal element $G \in \mathcal{E}$. By the previous step we know that this G is defined on the whole space H.

The proof is complete. □

A natural question whether there exists an extension theorem for mappings from a Hadamard space to itself has a negative answer; see [92, p. 119].

4.2 Resolvent of a nonexpansive mapping

We will now define resolvents of nonexpansive mappings. The reader may notice that a similar idea appeared already in the proof of Theorem 3.3.1.

Definition 4.2.1 (Resolvent of a nonexpansive mapping). Let $(\mathcal{H}, d)$ be a Hadamard space and $F\colon \mathcal{H} \to \mathcal{H}$ be nonexpansive. Given $x \in \mathcal{H}$ and $\lambda \in (0, \infty)$, the mapping

$$G_{x,\lambda}\colon y \mapsto \frac{1}{1+\lambda}x + \frac{\lambda}{1+\lambda}Fy, \quad y \in \mathcal{H}, \tag{4.2.5}$$

is a contraction with Lipschitz constant $\frac{\lambda}{1+\lambda}$ and hence has a unique fixed point, which will be denoted $R_\lambda x$. We put $R_0 x := x$ for every $x \in \mathcal{H}$. The mapping $x \mapsto R_\lambda x$ will be called the *resolvent* of F.

Recall that we have defined the resolvent of a convex lsc function (Definition 2.2.20) and denoted it by J_λ. In Hilbert spaces, they are both special cases of a resolvent of a maximal monotone operator; see Exercise 4.2. It is not surprising then that J_λ shares many properties with R_λ.

Lemma 4.2.2. *Let $F\colon \mathcal{H} \to \mathcal{H}$ be nonexpansive and $\lambda \in (0, \infty)$. Then the resolvent R_λ is firmly nonexpansive.*

Proof. First observe that R_λ is nonexpansive on account of the joint convexity of the metric in (1.2.4). Next, denote $u := (1-t)x + tR_\lambda x$ and $v := (1-t)y + tR_\lambda y$ for some $t \in (0,1)$ and $x, y \in \mathcal{H}$. Then,

$$R_\lambda x = (1-s)u + sFR_\lambda x, \quad R_\lambda y = (1-s)v + sFR_\lambda y,$$

where $s := \frac{\lambda - t\lambda}{1+\lambda-t\lambda}$ and, by the joint convexity of the metric (1.2.4), we have

$$\begin{aligned} d\left(R_\lambda x, R_\lambda y\right) &\le (1-s)d(u,v) + sd\left(FR_\lambda x, FR_\lambda y\right) \\ &\le (1-s)d(u,v) + sd\left(R_\lambda x, R_\lambda y\right), \end{aligned}$$

which along with (1.2.4) gives the desired conclusion. □

As an easy observation note that, for every $\lambda \in (0,\infty)$ and $x \in \mathcal{H}$, we have $R_\lambda x = x$ if and only if $Fx = x$. Therefore the fixed point set of a *nonexpansive* mapping F can be sought for as the fixed point set of the *firmly nonexpansive* mapping R_λ.

The following notation will be used at various places throughout the book.

Notation 4.2.3. Given a mapping $F\colon \mathcal{H} \to \mathcal{H}$, denote its nth-power by

$$F^{(n)} := F \circ \cdots \circ F, \quad n \in \mathbb{N},$$

where F appears n-times in the composition on the right-hand side. As a convention, we put $F^{(0)}x := x$ for every $x \in \mathcal{H}$.

Lemma 4.2.4. *Let $F\colon \mathcal{H} \to \mathcal{H}$ be a nonexpansive mapping. Then its resolvent satisfies*

$$d\left(x, R_\lambda x\right) \le \lambda d(x, Fx), \quad x \in \mathcal{H},$$

for every $\lambda \in (0, \infty)$.

Proof. Since $R_\lambda x$ is a fixed point of the contraction $G_{x,\lambda}$ from (4.2.5), it can be approximated by the Banach contraction principle. Therefore,

$$\begin{aligned} d\left(x, R_\lambda x\right) &\le \sum_{n=1}^{\infty} d\left(G_{x,\lambda}^{(n-1)}(x), G_{x,\lambda}^{(n)}(x)\right) \\ &\le d\left(x, G_{x,\lambda}(x)\right) \sum_{n=1}^{\infty} \left(\frac{\lambda}{1+\lambda}\right)^{n-1} \\ &\le (1+\lambda) d\left(x, G_{x,\lambda}(x)\right), \end{aligned}$$

and we are done because the right-hand side is equal to $\lambda d(x, Fx)$. □

Corollary 4.2.5. *Let $F\colon \mathcal{H} \to \mathcal{H}$ be a nonexpansive mapping. Then its resolvent satisfies*

$$d\left(R_\lambda^{(n)} x, x\right) \le n\lambda d(x, Fx), \quad x \in \mathcal{H},$$

for every $\lambda \in (0, \infty)$ and $n \in \mathbb{N}$.

Proof. Lemma 4.2.4 gives

$$d\left(R_\lambda^{(n)} x, x\right) \le \sum_{k=1}^{n} d\left(R_\lambda^{(k)} x, R_\lambda^{(k-1)} x\right) \le n d\left(R_\lambda x, x\right) \le n\lambda d(x, Fx),$$

which is exactly the statement. □

Proposition 4.2.6 (Resolvent identity). *Let $F\colon \mathcal{H} \to \mathcal{H}$ be a nonexpansive mapping. Then its resolvent satisfies*

$$R_\lambda x = R_\mu\left(\frac{\lambda-\mu}{\lambda} R_\lambda x + \frac{\mu}{\lambda} x\right), \quad x \in \mathcal{H},$$

for every $\lambda > \mu > 0$.

Proof. Denote $y := \frac{\lambda-\mu}{\lambda} R_\lambda x + \frac{\mu}{\lambda} x$ and recall that $R_\mu y$ is the unique point $z \in \mathcal{H}$ satisfying

$$d(y, z) = \frac{\mu}{1+\mu} d(y, Fz), \quad \text{and} \quad d(z, Fz) = \frac{1}{1+\mu} d(y, Fz).$$

It therefore suffices to show that these two equalities hold for $z := R_\lambda x$. By easy computations we have

$$d\left(y, R_\lambda x\right) = \frac{\mu}{1+\lambda} d\left(x, FR_\lambda x\right) = \frac{\mu}{1+\mu} d\left(y, FR_\lambda x\right),$$

which shows the first desired equality and the second one is analogous. □

Lemma 4.2.7. *Let $F\colon \mathcal{H} \to \mathcal{H}$ be nonexpansive and $x \in \mathcal{H}$. Then the mapping $\lambda \mapsto R_\lambda x$ is continuous on $[0, \infty)$.*

Proof. Lemma 4.2.4 gives $R_\lambda x \to x$ as $\lambda \to 0$. Continuity at a positive λ follows by the resolvent identity (Proposition 4.2.6) in a similar way as in the proof of Proposition 2.2.26. □

The following result is a counterpart to Theorem 2.2.25.

Theorem 4.2.8. *Let $F\colon \mathcal{H} \to \mathcal{H}$ be a nonexpansive mapping and $x \in \mathcal{H}$. If there exists $(\lambda_n) \subset (0, \infty)$ such that $\lambda_n \to \infty$ and the sequence $(R_{\lambda_n} x)_{n\in\mathbb{N}}$ is bounded, then* Fix F *is nonempty and*

$$\lim_{\lambda\to\infty} R_\lambda x = P_{\operatorname{Fix} F}(x)\,. \tag{4.2.6}$$

Conversely, if Fix $F \neq \emptyset$, *then $(R_\lambda x)_{\lambda\in(0,\infty)}$ is bounded.*

Proof. To simplify our notation, put $x_\lambda := R_\lambda(x)$ for each $\lambda \in (0, \infty)$. Fix now $0 < \mu < \lambda$ and let

$$\triangle\left(\overline{x}, \overline{Fx_\lambda}, \overline{Fx_\mu}\right) \subset \mathbb{R}^2$$

be a comparison triangle for $\triangle(x, Fx_\lambda, Fx_\mu)$. We have

$$\left\|\overline{Fx_\lambda} - \overline{Fx_\mu}\right\| = d\left(Fx_\lambda, Fx_\mu\right) \le d\left(x_\lambda, x_\mu\right) \le \left\|\overline{x_\lambda} - \overline{x_\mu}\right\| ,$$

where $\overline{x_\lambda}$ and $\overline{x_\mu}$ are the comparison points for x_λ and x_μ, respectively, in the comparison triangle $\triangle(\overline{x}, \overline{Fx_\lambda}, \overline{Fx_\mu})$. Without loss of generality we can assume $\overline{x} = 0 \in \mathbb{R}^2$. From this and the fact that $\overline{x_\lambda} = \frac{\lambda}{1+\lambda}\overline{Fx_\lambda}$ and $\overline{x_\mu} = \frac{\mu}{1+\mu}\overline{Fx_\mu}$ we further obtain

$$\left\langle \frac{1+\lambda}{\lambda}\overline{x_\lambda} - \frac{1+\mu}{\mu}\overline{x_\mu}, \frac{1+\lambda}{\lambda}\overline{x_\lambda} - \frac{1+\mu}{\mu}\overline{x_\mu} \right\rangle \le \left\|\overline{x_\lambda} - \overline{x_\mu}\right\|^2 .$$

A simple computation yields

$$\begin{aligned}\left(\frac{1+\lambda}{\lambda} - \frac{1+\mu}{\mu}\right)^2 \left\|\overline{x_\mu}\right\|^2 + \left(\frac{(1+\lambda)^2}{\lambda^2} - 1\right)\left\|\overline{x_\lambda} - \overline{x_\mu}\right\|^2 \\ \le 2\left(\frac{1+\mu}{\mu} - \frac{1+\lambda}{\lambda}\right)\frac{1+\lambda}{\lambda}\left\langle \overline{x_\mu}, \overline{x_\lambda} - \overline{x_\mu}\right\rangle .\end{aligned}$$

Consequently,

$$\left\langle \overline{x_\mu}, \overline{x_\lambda} - \overline{x_\mu}\right\rangle \ge 0\,.$$

Since

$$\left\|\overline{x_\lambda}\right\|^2 = \left\|\overline{x_\mu}\right\|^2 + \left\|\overline{x_\lambda} - \overline{x_\mu}\right\|^2 + 2\left\langle \overline{x_\mu}, \overline{x_\lambda} - \overline{x_\mu}\right\rangle ,$$

we have

$$\left\|\overline{x_\mu}\right\| \le \left\|\overline{x_\lambda}\right\| ,$$

and,

$$d\left(x_\lambda, x_\mu\right)^2 \le \left\|\overline{x_\lambda} - \overline{x_\mu}\right\|^2 \le \left\|\overline{x_\lambda} - \overline{x}\right\|^2 - \left\|\overline{x_\mu} - \overline{x}\right\|^2 . \tag{4.2.7}$$

The monotonicity of $\lambda \mapsto \|\overline{x_\lambda}\|$ and the boundedness of $(\|\overline{x_{\lambda_n}}\|)_n$ together give that $(\|\overline{x_\lambda}\|)_{\lambda\in(0,\infty)}$ is bounded. Consequently, there exists $l \in [0,\infty)$ such that $\|\overline{x_\lambda}\| \to l$ as $\lambda \to \infty$. Inequality (4.2.7) therefore implies

$$d\left(x_\lambda, x_\mu\right)^2 \to 0, \quad \text{as } \lambda, \mu \to \infty.$$

Let $z \in \mathcal{H}$ be the limit point of (x_λ). By continuity we have

$$d(z, Fz) = \lim_{\lambda\to\infty} d\left(x_\lambda, Fx_\lambda\right) = \lim_{\lambda\to\infty} \frac{1}{1+\lambda} d\left(x, Fx_\lambda\right) = 0,$$

which means that $z \in \operatorname{Fix} F$.

We will now show $z = P_{\operatorname{Fix} F}(x)$. Let $p \in \operatorname{Fix} F$ and repeat the above argument with the triangle $\triangle(x, p, F(x_\mu))$. We obtain

$$d(x,p)^2 = \|\overline{x} - \overline{p}\|^2 \ge \left\|\overline{x} - \overline{x_\mu}\right\|^2 + \left\|\overline{x_\mu} - \overline{p}\right\|^2 \ge d\left(x, x_\mu\right)^2 + d\left(x_\mu, p\right)^2 ,$$

and after taking the limit $\mu \to \infty$, we have

$$d(x,p)^2 \ge d(x,z)^2 + d(z,p)^2 ,$$

which completes the proof that $z = P_{\operatorname{Fix} F}(x)$.

Finally, it is easy to see that if $\operatorname{Fix} F \neq \emptyset$, then $(x_\lambda)_{\lambda\in(0,\infty)}$ is bounded. □

4.3 Strongly continuous semigroup

We will now show that applying the resolvent R_λ of a nonexpansive mapping $F\colon \mathcal{H} \to \mathcal{H}$ iteratively and decreasing the step size λ generates a semigroup of nonexpansive mappings (T_t). The following two technical results are needed in the proof of Theorem 4.3.3.

Lemma 4.3.1. *Let $F\colon \mathcal{H} \to \mathcal{H}$ be a nonexpansive mapping. If $\lambda > \mu > 0$ and $m, n \in \mathbb{N}$, then the resolvent satisfies*

$$\begin{aligned} d\left(R_\mu^{(n)}x, R_\lambda^{(m)}x\right) \le{}& \sum_{k=0}^{m-1} \left(\frac{\mu}{\lambda}\right)^k \left(\frac{\lambda-\mu}{\lambda}\right)^{n-k} \binom{n}{k} d\left(R_\lambda^{(m-k)}x, x\right) \\ &+ \sum_{k=m}^{n} \left(\frac{\mu}{\lambda}\right)^m \left(\frac{\lambda-\mu}{\lambda}\right)^{k-m} \binom{k-1}{m-1} d\left(R_\mu^{(n-k)}x, x\right), \end{aligned}$$

whenever $m \le n$, whereas

$$d\left(R_\mu^{(n)}x, R_\lambda^{(m)}x\right) \le \sum_{k=0}^{n} \left(\frac{\mu}{\lambda}\right)^k \left(\frac{\lambda-\mu}{\lambda}\right)^{n-k} \binom{n}{k} d\left(R_\lambda^{(m-k)}x, x\right) ,$$

provided $m > n$.

Proof. To derive the above inequalities, one can mimic the original Crandall–Liggett proof [68, Lemma 1.3]. □

Lemma 4.3.2. *Let $\alpha \in (0,1)$ and $m, n \in \mathbb{N}$ with $m \le n$. Then:*
(i) $\sum_{k=0}^{m-1} \binom{n}{k} \alpha^k (1-\alpha)^{n-k}(m-k) \le \sqrt{(n\alpha - m)^2 + n\alpha(1-\alpha)}$,
(ii) $\sum_{k=m}^{n} \binom{k-1}{m-1} \alpha^m (1-\alpha)^{k-m}(n-k) \le \sqrt{\frac{m(1-\alpha)}{\alpha^2} + (\frac{m(1-\alpha)}{\alpha} + m - n)^2}$.

Proof. See Exercise 4.3. □

We can now state the main result of the current section.

Theorem 4.3.3 (Existence of the semigroup). *Let $(\mathcal{H}, d)$ be a Hadamard space and $F\colon \mathcal{H} \to \mathcal{H}$ be a nonexpansive mapping. Then the limit*

$$T_t x := \lim_{n\to\infty} R_{\frac{t}{n}}^{(n)} x\,, \quad x \in \mathcal{H}\,, \tag{4.3.8}$$

exists and is uniform with respect to t on each bounded subinterval of $[0,\infty)$. Moreover, the family $(T_t)_t$ is a strongly continuous semigroup of nonexpansive mappings, that is,
(i) $\lim_{t\to 0+} T_t x = x$,
(ii) $T_t(T_s x) = T_{s+t} x$, *for every* $s, t \in [0,\infty)$,
(iii) $d(T_t x, T_t y) \le d(x,y)$, *for each* $t \in [0,\infty)$,
for every $x, y \in \mathcal{H}$. Finally, we have the estimates

$$d\left(T_t x, T_s x\right) \le 2d(x, Fx)|t-s|\,, \quad x \in \mathcal{H}, \tag{4.3.9}$$

for every $s, t \in [0,\infty)$, and

$$d\left(T_t x, R_{\frac{t}{n}}^{(n)} x\right) \le \frac{2t}{\sqrt{n}} d(x, Fx)\,, \quad x \in \mathcal{H}\,, \tag{4.3.10}$$

for every $t \in [0,\infty)$ and $n \in \mathbb{N}$.

Proof. We divide the proof into several steps. Step 1: Choose $\lambda \ge \mu > 0$ and $m, n \in \mathbb{N}$ with $m \le n$. Applying Lemma 4.3.1, Corollary 4.2.5 and Lemma 4.3.2, we obtain

$$\begin{aligned} d\left(R_\mu^{(n)} x, R_\lambda^{(m)} x\right) &\le \lambda \sum_{k=1}^{m-1} \left(\frac{\mu}{\lambda}\right)^k \left(\frac{\lambda-\mu}{\lambda}\right)^{n-k} \binom{n}{k} (m-k) d(x, Fx) \\ &\quad + \mu \sum_{k=m}^{n} \left(\frac{\mu}{\lambda}\right)^m \left(\frac{\lambda-\mu}{\lambda}\right)^{k-m} \binom{k-1}{m-1} (n-k) d(x, Fx) \\ &\le \lambda d(x, Fx) \sqrt{\left(n\frac{\mu}{\lambda} - m\right)^2 + n\frac{\mu}{\lambda}\frac{\lambda-\mu}{\lambda}} \\ &\quad + \mu d(x, Fx) \sqrt{\frac{\lambda^2}{\mu^2}\frac{\lambda-\mu}{\lambda} m + \left(\frac{\lambda}{\mu}\frac{\lambda-\mu}{\lambda} m + m - n\right)^2} \end{aligned}$$

$$= d(x, Fx)\sqrt{(n\mu - m\lambda)^2 + n\mu(\lambda - \mu)}$$

$$+ d(x, Fx)\sqrt{m\lambda(\lambda - \mu) + (m\lambda - n\mu)^2}\,. \tag{4.3.11}$$

If we put $\mu := \frac{t}{n}$ and $\lambda := \frac{t}{m}$, the above inequality reads

$$d\left(R_{\frac{t}{n}}^{(n)}x, R_{\frac{t}{m}}^{(m)}x\right) \le 2td(x, Fx)\sqrt{\frac{1}{m} - \frac{1}{n}}\,.$$

We immediately get the existence of the limit (4.3.8), uniformly in t on bounded intervals, as well as the error estimate (4.3.10). Consequently also (i) holds with the help of Lemma 4.2.7.

Step 2: In order to show (4.3.9) for $0 < s \le t$, it suffices to put $\mu := \frac{s}{n}$ and $\lambda := \frac{t}{n}$ along with $m = n$ in (4.3.11) and take the limit. The cases $s = 0$ or $t = 0$ then follow by (i). It is easy to see that T_t is a nonexpansive mapping.

Step 3: It remains to prove the semigroup property in (ii). Let us first show that $T_{kt} = T_t^{(k)}$, where $t \in (0, \infty)$ and $k \in \mathbb{N}$. We proceed by induction on k. For $k = 2$, we have

$$d\left(T_t^{(2)}x, R_{\frac{t}{n}}^{(2n)}x\right) \le d\left(T_tT_tx, R_{\frac{t}{n}}^{(n)}T_tx\right) + d\left(R_{\frac{t}{n}}^{(n)}T_tx, R_{\frac{t}{n}}^{(n)}R_{\frac{t}{n}}^{(n)}x\right),$$

and the right-hand side tends to 0 as $n \to \infty$. Therefore,

$$T_t^{(2)}x = \lim_{n\to\infty} R_{\frac{t}{n}}^{(2n)}x = \lim_{n\to\infty} R_{\frac{2t}{2n}}^{(2n)}x = T_{2t}x\,.$$

Choose now $k \in \mathbb{N}$ with $k \ge 3$ and assume the claim holds for $k - 1$. Then

$$d\left(T_t^{(k)}x, R_{\frac{t}{n}}^{(kn)}x\right) \le d\left(T_tT_t^{(k-1)}x, R_{\frac{t}{n}}^{(n)}T_t^{(k-1)}x\right)$$
$$+ d\left(R_{\frac{t}{n}}^{(n)}T_t^{(k-1)}x, R_{\frac{t}{n}}^{(n)}R_{\frac{t}{n}}^{(kn-n)}x\right),$$

and again the right-hand side tends to 0 as $n \to \infty$ by our assumption. We hence have

$$T_t^{(k)}x = \lim_{n\to\infty} R_{\frac{t}{n}}^{(kn)}x = \lim_{n\to\infty} R_{\frac{kt}{kn}}^{(kn)}x = T_{kt}x\,.$$

Consequently, we obtain $T_tT_s = T_{s+t}$ for s, t rational, that is, $s := \frac{k}{l}$ and $t := \frac{p}{q}$ for some $k, l, p, q \in \mathbb{N}$. Indeed,

$$T_{\frac{k}{l}+\frac{p}{q}} = T_{\frac{kq+lp}{lq}} = T_{\frac{1}{lq}}^{(kq+lp)} = T_{\frac{1}{lq}}^{(kq)}T_{\frac{1}{lq}}^{(lp)} = T_{\frac{k}{l}}T_{\frac{p}{q}}\,.$$

Finally, by the continuity of $t \mapsto T_tx$ we get the desired semigroup property for every $s, t \in [0, \infty)$. The proof is complete. □

By Lemma 4.2.4 we have

$$d\left(x, R_{\frac{t}{n}}^{(n)}x\right) \le \sum_{j=0}^{n-1} d\left(R_{\frac{t}{n}}^{(j)}x, R_{\frac{t}{n}}^{(j+1)}x\right) \le nd\left(x, R_{\frac{t}{n}}x\right) = td(x, Fx)\,,$$

and taking the limit $n \to \infty$ gives

$$d\left(x, T_t x\right) \leq t d(x, Fx)\,, \tag{4.3.12}$$

for every $t \in [0, \infty)$. Before stating the following theorem, recall that the weak convergence of a curve was defined in Definition 3.1.9.

Theorem 4.3.4 (Asymptotic behavior of the semigroup). *Let $F \colon \mathcal{H} \to \mathcal{H}$ be a nonexpansive mapping with at least one fixed point and $x \in \mathcal{H}$. Then $T_t x$ weakly converges to a fixed point of F as $t \to \infty$.*

Proof. Step 1: First observe that

$$d\left(R_\lambda x, FR_\lambda x\right) = \frac{1}{\lambda} d\left(x, R_\lambda x\right) \leq d(x, Fx)\,,$$

by Lemma 4.2.4. We hence have

$$d(x, Fx) \geq d\left(R_{\frac{t}{n}} x, FR_{\frac{t}{n}} x\right) \geq d\left(R_{\frac{t}{n}}^{(n)} x, FR_{\frac{t}{n}}^{(n)} x\right)\,,$$

and after taking the limit $n \to \infty$, also

$$d(x, Fx) \geq d\left(T_t x, FT_t x\right)\,.$$

The semigroup property implies

$$d\left(T_s x, FT_s x\right) \geq d\left(T_t x, FT_t x\right)\,,$$

whenever $s \leq t$ and therefore the limit

$$\lim_{t\to\infty} d\left(T_t x, FT_t x\right) \tag{4.3.13}$$

exists.

Step 2: We will now show that this limit actually equals to 0. Let $0 \leq s \leq t$. Then inequality (4.3.12) gives

$$\begin{aligned} d\left(T_s x, T_t x\right) &\leq \sum_{j=0}^{n-1} d\left(T_{s+\frac{j}{n}(t-s)} x, T_{s+\frac{j+1}{n}(t-s)} x\right) \\ &\leq \frac{t-s}{n} \sum_{j=0}^{n-1} d\left(T_{s+\frac{j}{n}(t-s)} x, FT_{s+\frac{j}{n}(t-s)} x\right)\,, \end{aligned}$$

and after sending $n \to \infty$, we obtain

$$d\left(T_s x, T_t x\right) \leq \int_s^t d\left(T_r x, FT_r x\right) \mathrm{d}r\,. \tag{4.3.14}$$

Next we prove that

$$\lim_{t\to\infty} d\left(T_t x, FT_t x\right) \le \frac{1}{h} \lim_{t\to\infty} d\left(T_{t+h}x, T_t x\right) . \tag{4.3.15}$$

To this end, we repeatedly use the inequality

$$d\left(FR_\lambda^{(n)}x, R_\lambda^{(n-k+1)}x\right) \le \frac{1}{1+\lambda} d\left(FR_\lambda^{(n)}x, R_\lambda^{(n-k)}x\right) + \frac{\lambda}{1+\lambda} d\left(R_\lambda^{(n)}x, R_\lambda^{(n-k+1)}x\right) ,$$

valid for each $1 \le k \le n$, to obtain

$$d\left(FR_\lambda^{(n)}x, R_\lambda^{(n)}x\right) \le \frac{1}{(1+\lambda)^n} d\left(FR_\lambda^{(n)}x, x\right) + \lambda \sum_{j=1}^{n} \frac{1}{(1+\lambda)^j} d\left(R_\lambda^{(n)}x, R_\lambda^{(n-j+1)}x\right) .$$

Put now $\lambda := \frac{t}{n}$ and take the limit $n \to \infty$. One arrives at

$$d\left(T_t x, FT_t x\right) \le \exp(-t) d\left(FT_t x, x\right) + \int_0^t \exp(-r) d\left(T_t x, T_{t-r}x\right) \mathrm{d}r .$$

Applying inequality (4.3.14) and an elementary calculation we arrive at

$$\exp(t) d\left(T_t x, FT_t x\right) \le d\left(FT_t x, x\right) + \int_0^t \left(\exp(r) - 1\right) d\left(T_r x, FT_r x\right) \mathrm{d}r ,$$

or

$$\left(\exp(t) - 1\right) d\left(T_t x, FT_t x\right) \le d\left(T_t x, x\right) + \int_0^t \left(\exp(r) - 1\right) d\left(T_r x, FT_r x\right) \mathrm{d}r .$$

Replacing t by h and then x by $T_t x$ gives

$$\begin{aligned}\left(\exp(h) - 1\right) d\left(T_{t+h}x, FT_{t+h}x\right) \le & \int_t^{t+h} \left(\exp(r-t) - 1\right) d\left(T_r x, FT_r x\right) \mathrm{d}r \\ & + d\left(T_{t+h}x, T_t x\right) .\end{aligned}$$

By an easy calculation we obtain

$$\begin{aligned} d\left(T_{t+h}x, T_t x\right) \ge & \left(\exp(h) - 1\right) \left[d\left(T_{t+h}x, FT_{t+h}x\right) - d\left(T_t x, FT_t x\right)\right] \\ & + h d\left(T_t x, FT_t x\right) ,\end{aligned}$$

which proves (4.3.15). Then (4.3.15) and (4.3.14) yield

$$\begin{aligned} \lim_{t\to\infty} d\left(T_t x, FT_t x\right) & \le \limsup_{h\to\infty} \frac{1}{h} d\left(T_h x, x\right) \\ & \le \lim_{h\to\infty} \frac{1}{h} \int_0^h d\left(T_r x, FT_r x\right) \mathrm{d}r \\ & = \lim_{t\to\infty} d\left(T_t x, FT_t x\right) ,\end{aligned}$$

and thus

$$\lim_{t\to\infty} d\left(T_t x, FT_t x\right) = \limsup_{h\to\infty} \frac{1}{h} d\left(T_h x, x\right). \tag{4.3.16}$$

Let now $y \in \mathcal{H}$. Then

$$\begin{aligned}\limsup_{h\to\infty} \frac{d\left(T_h x, x\right)}{h} &\leq \limsup_{h\to\infty} \frac{1}{h}\left[d\left(T_h x, T_h y\right) + d\left(T_h y, y\right) + d\left(y, x\right)\right] \\ &\leq \limsup_{h\to\infty} \frac{d\left(T_h y, y\right)}{h},\end{aligned}$$

and since y was arbitrary, the value on the left hand side is independent of x. Consequently, by virtue of (4.3.16), the limit in (4.3.13) is independent of x and is therefore equal to 0, because one may choose $x \in \operatorname{Fix} F$.

Step 3: To finish the proof, choose a sequence $t_n \to \infty$ and put $x_n := T_{t_n} x_0$. Since T_t is nonexpansive, we know that (x_n) is Fejér monotone with respect to $\operatorname{Fix} F$. In particular, the sequence (x_n) is bounded and has therefore a weak cluster point $z \in \mathcal{H}$. In order to apply Proposition 3.2.6, it suffices to show that $z \in \operatorname{Fix} F$. We easily get

$$\begin{aligned}\limsup_{n\to\infty} d\left(Fz, x_n\right) &\leq \limsup_{n\to\infty} d\left(Fz, Fx_n\right) + \limsup_{n\to\infty} d\left(Fx_n, x_n\right) \\ &\leq \limsup_{n\to\infty} d\left(z, x_n\right),\end{aligned}$$

which by the uniqueness of the weak limit gives $z = Fz$. Here we used, of course, the fact that the limit in (4.3.13) is 0.

It is easy to see that z is independent of the choice of the sequence (t_n) and therefore $T_t x_0 \xrightarrow{w} z$. □

As noted in [24, Remark, p. 7], there exists a counterexample in Hilbert space showing that the convergence in Theorem 4.3.4 is not strong in general.

We finish this chapter by a rather technical result which will be needed in Section 5.3.

Lemma 4.3.5. *Let $F\colon \mathcal{H} \to \mathcal{H}$ be a nonexpansive mapping. Then its semigroup satisfies*

$$d\left(T_t x, F^{(k)} x\right) \leq 2d(x, Fx)\sqrt{(k-t)^2 + t}, \quad x \in \mathcal{H},$$

for every $t \in [0,\infty)$ and $k \in \mathbb{N}_0$.

Proof. Step 1: Choose $n \in \mathbb{N}$. The convexity of the distance function gives

$$d\left(F^{(k)}x, R^{(n)}_{\frac{t}{n}} x\right)^2 \leq \frac{1}{1+\frac{t}{n}} d\left(F^{(k)}x, R^{(n-1)}_{\frac{t}{n}} x\right)^2 + \frac{\frac{t}{n}}{1+\frac{t}{n}} d\left(F^{(k)}x, FR^{(n)}_{\frac{t}{n}} x\right)^2,$$

and repeating the argument n-times, we obtain

$$\begin{aligned} d\left(F^{(k)}x, R^{(n)}_{\frac{t}{n}}x\right)^2 &\le \left(1+\frac{t}{n}\right)^{-n} d\left(F^{(k)}x, x\right)^2 \\ &\quad + \frac{t}{n}\sum_{j=1}^{n}\left(1+\frac{t}{n}\right)^{-j} d\left(F^{(k)}x, FR^{(n-j+1)}_{\frac{t}{n}}x\right)^2 \\ &\le \left(1+\frac{t}{n}\right)^{-n} k^2 d(Fx,x)^2 \\ &\quad + \frac{t}{n}\sum_{j=1}^{n}\left(1+\frac{t}{n}\right)^{-j} d\left(F^{(k-1)}x, R^{(n-j+1)}_{\frac{t}{n}}x\right)^2 \\ &= \left(1+\frac{t}{n}\right)^{-n} k^2 d(Fx,x)^2 \\ &\quad + \frac{t}{n}\sum_{j=1}^{n}\left(1+\frac{t}{n}\right)^{-n-j} d\left(F^{(k-1)}x, R^{(j)}_{\frac{t}{n}}x\right)^2 . \end{aligned}$$

If we define the functions

$$\varphi_n(s) := \sum_{j=1}^{n}\left(1+\frac{s}{n}\right)^{-n-j} \chi_{\left(\frac{(j-1)t}{n}, \frac{jt}{n}\right]}(s) ,$$

and,

$$\psi_n(s) := \sum_{j=1}^{n} d\left(F^{(k-1)}x, R^{(j)}_{\frac{s}{n}}x\right)^2 \chi_{\left(\frac{(j-1)t}{n}, \frac{jt}{n}\right]}(s) ,$$

where $s \in [0,t]$ and $n \in \mathbb{N}$, the above inequality reads

$$d\left(F^{(k)}x, R^{(n)}_{\frac{t}{n}}x\right)^2 \le \left(1+\frac{t}{n}\right)^{-n} k^2 d\left(Fx,x\right)^2 + \int_0^t \varphi_n(s)\psi_n(s)\,\mathrm{d}s .$$

Step 2: We claim that

$$\lim_{n\to\infty} \varphi_n(s) = \exp(s-t) \tag{4.3.17}$$

and,

$$\lim_{n\to\infty} \psi_n(s) = d\left(F^{(k-1)}x, T_s x\right)^2 ,$$

for every $s \in (0,t]$ and that $\sup_{n\in\mathbb{N}, s\in[0,t]} |\varphi_n(s)|\,|\psi_n(s)| < \infty$. Then we could use the dominated convergence theorem and conclude

$$d\left(F^{(k)}x, T_t x\right)^2 \le k^2 \exp(-t) d(x, Fx) + \int_0^t \exp(s-t) d\left(F^{(k-1)}x, T_s x\right)^2 \mathrm{d}s . \tag{4.3.18}$$

Let us prove the claim. For every $n \in \mathbb{N}$ and $s \in (0, t]$ there exists a unique $\nu = \nu(n, s) \in \mathbb{N}$ such that $(\nu - 1)t < sn \leq \nu t$. As a matter of fact, it satisfies $\nu \leq n$. Inequality (4.3.11) with $m := \nu$, $\lambda := \frac{t}{n}$, $n := l$ and $\mu := \frac{s}{l}$ reads

$$d\left(R_{\frac{s}{l}}^{(l)}x, R_{\frac{t}{n}}^{(\nu)}x\right) \leq d(x, Fx)\sqrt{\left(s - \frac{\nu t}{n}\right)^2 + s\left(\frac{t}{n} - \frac{s}{l}\right)}$$
$$+ d(x, Fx)\sqrt{\frac{\nu t}{n}\left(\frac{t}{n} - \frac{s}{l}\right) + \left(s - \frac{\nu t}{n}\right)^2},$$

and, since we assume $\nu \leq l$, we can take the limit $l \to \infty$ to arrive at

$$d\left(T_s x, R_{\frac{t}{n}}^{(\nu)}x\right) \leq d(x, Fx)\left(\sqrt{\left(s - \frac{\nu t}{n}\right)^2 + \frac{st}{n}} + \sqrt{\frac{\nu t^2}{n^2} + \left(s - \frac{\nu t}{n}\right)^2}\right).$$

Since

$$\left|\psi_n(s) - d\left(F^{(k-1)}x, T_s x\right)^2\right| = \left|d\left(F^{(k-1)}x, R_{\frac{t}{n}}^{(\nu)}x\right) + d\left(F^{(k-1)}x, T_s x\right)\right|$$
$$\cdot \left|d\left(F^{(k-1)}x, R_{\frac{t}{n}}^{(\nu)}x\right) - d\left(F^{(k-1)}x, T_s x\right)\right|,$$

and the set $\{R_{\frac{t}{n}}^{(i)}x \colon n, i \in \mathbb{N},\ i \leq n\}$ is bounded, the above inequality gives

$$\lim_{n\to\infty} \psi_n(s) = d\left(F^{(k-1)}x, T_s x\right)^2, \quad s \in (0, t],$$

as well as that (ψ_n) is a uniformly bounded sequence. The analogous claims about φ_n are an easy exercise and are omitted. We now justified (4.3.18).

Step 3: If $d(x, Fx) = 0$, the lemma holds trivially. Let us therefore assume $d(x, Fx) > 0$ and put

$$\xi_k(s) := \frac{d\left(F^{(k)}x, T_s x\right)}{2d(x, Fx)}, \quad s \in [0, \infty),$$

where $k \in \mathbb{N}_0$. Then (4.3.18) becomes

$$\xi_k(t)^2 \leq \frac{1}{2}\exp(-t)k^2 + \int_0^t \exp(s - t)\xi_{k-1}(s)^2\, \mathrm{d}s, \tag{4.3.19}$$

and we need to show

$$\xi_k(t)^2 \leq (k - t)^2 + t,$$

to complete the proof. We will proceed by induction on k. The case $k = 0$ is a consequence of (4.3.9). Fix now $k \in \mathbb{N}$ and assume the desired inequality holds for $k - 1$. Then by (4.3.19) we have

$$\xi_k(t)^2 \leq \frac{1}{2}\exp(-t)k^2 + \int_0^t \exp(s - t)\left[(k - 1 - s)^2 + s\right] \mathrm{d}s,$$

and it suffices to show that

$$\frac{1}{2}k^2 + \int_0^t \exp(s)\left[(k-1-s)^2 + s\right] \mathrm{d}s = \exp(t)\left[(k-t)^2 + t\right],$$

which is obviously the case, for the expressions on either side coincide at $t = 0$ and their derivatives with respect to t are equal. This finishes the proof. □

Exercises

Exercise 4.1. Show that if we assumed F to be $(1+\varepsilon)$-Lipschitz in Theorem 3.3.1, for some $\varepsilon > 0$, instead of 1-Lipschitz, the statement fails.
Hint. Let $B \subset \ell_2$ be the closed unit ball of the Hilbert space ℓ_2. Choose $\lambda > 0$ and consider the mapping $F: B \to B$ defined by

$$F: (x_1, x_2, x_3, \dots) \mapsto \left(\frac{1 - \|x\|}{\lambda}, x_1, x_2, \dots\right),$$

for every $x = (x_1, x_2, x_3, \dots) \in \ell_2$. □

Exercise 4.2. Let H be a Hilbert space. Show that the resolvent of a convex lsc function $f: H \to (-\infty, \infty]$ satisfies $J_\lambda = (I + \lambda\partial f)^{-1}$. Furthermore, show that the resolvent of a nonexpansive mapping $F: H \to H$ satisfies $R_\lambda = (I + \lambda(I - F))^{-1}$.

Exercise 4.3. Prove Lemma 4.3.2.
Hint. See [68, Lemma 1.4]. □

Exercise 4.4. Give the missing details in the proof of Lemma 4.3.5, that is, prove (4.3.17) and show that the sequence (φ_n) is uniformly bounded.

Bibliographical remarks

In Section 4.1, we follow the original paper [127] by U. Lang and V. Schroeder. There have been many important follow-up results, for instance [53, 126, 128, 141, 149]. These developments were motivated by the famous papers [113] by M. Kirszbraun and [200] by F. Valentine. A beautiful proof of the classical result is due to S. Reich and S. Simons [176]. Interestingly, a similar extension theorem for *firmly* nonexpansive mappings in Hilbert spaces was proved by H. Bauschke [30] and H. Bauschke and X. Wang [33].

The material of Sections 4.2 and 4.3 comes mostly from [189], where the author developed a Hadamard space analog of the theory due to M. Crandall and T. Liggett [68].

Formula (4.3.8) was in a nonlinear context used already in [173, Theorem 8.1]. Theorem 4.2.8 is due to W. A. Kirk [112, Theorem 26]. Theorem 4.3.4 was proved in [21] mimicking a Hilbert ball result of S. Reich [171]. In this connection, see also the paper [24] by J.-B. Baillon, R. E. Bruck and S. Reich.

We would like to mention a recent monograph by A. Zaslavski and S. Reich [178], where the interested reader can find a lot of information on nonexpansive mappings in linear and hyperbolic spaces as well as other aspects of nonlinear analysis discussed in our book.

5 Gradient flow of a convex functional

Let $(\mathcal{H}, d)$ be a Hadamard space and $f\colon \mathcal{H} \to (-\infty, \infty]$ be a convex lsc function. It turns outs that the resolvent of f generates a semigroup of nonexpansive mappings (S_t) which for a given point $x \in \overline{\operatorname{dom} f}$ represents a curve $x_t := S_t x$ moving in the direction of the steepest descent of f. This is called a *gradient flow.*

In the present chapter, we first prove the existence of a gradient flow in Hadamard spaces and show that the properties of a gradient flow in a Hilbert space carry over into our nonlinear setting. In Section 5.2 we will see that if a sequence of functions converges in the sense of Mosco, the corresponding resolvents (resp. semigroups) converge pointwise to the resolvent (resp. semigroup) of the limit function. Finally, in Section 5.3 we derive a nonlinear version of the celebrated Lie–Trotter–Kato formula.

We start by recalling the classical theory of gradient flows. Let H be a Hilbert space. Given a smooth convex function $f\colon H \to \mathbb{R}$ and $x_0 \in H$, consider the following parabolic problem

$$\begin{aligned} \dot{x}(t) &= -\nabla f(x(t)), \quad t \in (0, \infty), \\ x(0) &= x_0, \end{aligned} \tag{5.0.1}$$

for an unknown curve $x\colon [0, \infty) \to H$. In other words, one looks for a curve starting at x_0 which moves in the direction of the steepest descent of the function f. A solution $x\colon [0, \infty) \to H$ is then called a *gradient flow* governed by the functional f. If we assume f be only convex lsc, we can study a similar parabolic problem after replacing the gradient ∇f by the subgradient ∂f. More precisely, let $f\colon H \to (-\infty, \infty]$ be a convex lsc function and $x_0 \in H$. Then the problem

$$\begin{aligned} \dot{x}(t) &\in -\partial f(x(t)), \quad t \in (0, \infty), \\ x(0) &= x_0, \end{aligned} \tag{5.0.2}$$

for an unknown curve $x\colon [0, \infty) \to H$ is a natural generalization of (5.0.1) and has been well understood for quite some time.

While a generalization of (5.0.2) into Riemannian manifolds is rather straightforward, it is less obvious how to define gradient flows for convex lsc functionals in Hadamard spaces. The main obstacle is that there is no notion of a subdifferential in Hadamard spaces and hence we must adopt a completely different approach. Quite awesomely, we will in the present chapter arrive at a very satisfactory definition of a gradient flow, which is fully equivalent to (5.0.2) in classical settings and behaves equally well.

5.1 Gradient flow semigroup

Let $(\mathcal{H}, d)$ be a Hadamard space and $f\colon \mathcal{H} \to (-\infty, \infty]$ be a convex lsc function. Given a point $x_0 \in \mathcal{H}$, we would like to find a curve $x\colon [0,\infty) \to \mathcal{H}$ with $x(0) = x_0$ which at each point moves in the direction of the steepest descent of the function f. To this end, we define a *slope* of a function and derive its properties prior to stating the main result on the existence of a gradient flow in Theorem 5.1.6.

Slope of a function

The notion of a slope is to partially compensate for the lack of a subdifferential of a convex lsc function, but in contrast to the subdifferential, the slope is a nonnegative real *number*.

Definition 5.1.1 (Slope of a function). Let $f\colon \mathcal{H} \to (-\infty, \infty]$ be a convex lsc function and $x \in \operatorname{dom} f$. Define the *slope* of f at x as

$$|\partial f|(x) := \limsup_{y \to x} \frac{\max\{f(x) - f(y), 0\}}{d(x,y)},$$

and $\operatorname{dom} |\partial f| := \{x \in \mathcal{H}\colon |\partial f|(x) < \infty\}$. If $f(x) = \infty$, we set $|\partial f|(x) := \infty$.

Note that if H is a Hilbert space and $f\colon H \to \mathbb{R}$ is smooth, then $\|\nabla f(x)\| = |\partial f|(x)$ for each $x \in H$.

The definition of a slope has a local character. The following lemma shows that it can be equivalently obtained as a global supremum. As a consequence we get the lower semicontinuity of the function $x \mapsto |\partial f|(x)$ on $\operatorname{dom} |\partial f|(x)$.

Lemma 5.1.2. *Let $f\colon \mathcal{H} \to (-\infty, \infty]$ be a convex lsc function. Then*

$$|\partial f|(x) = \sup_{y \in \mathcal{H}\setminus\{x\}} \frac{\max\{f(x) - f(y), 0\}}{d(x,y)}, \quad x \in \operatorname{dom} f.$$

Furthermore, the slope of f is lsc, that is,

$$|\partial f|(x) \le \liminf_{n\to\infty} |\partial f|\,(x_n),$$

whenever $x_n \to x$ and $x \in \operatorname{dom} |\partial f|$.

Proof. It is immediate that

$$|\partial f|(x) \le \sup_{y \in \mathcal{H}\setminus\{x\}} \frac{\max\{f(x) - f(y), 0\}}{d(x,y)}.$$

To show the converse inequality, we may assume that $x \in \operatorname{dom} |\partial f|$ and $y \in \mathcal{H}$ satisfy $f(x) - f(y) > 0$. If we denote $\gamma\colon [0,1] \to \mathcal{H}$ the geodesic from x to y, then by convexity

$$\frac{f(x) - f(\gamma_t)}{d(x, \gamma_t)} \geq \frac{f(x) - f(y)}{d(x, y)},$$

and hence,

$$|\partial f|(x) \geq \limsup_{t \to 0+} \frac{f(x) - f(\gamma_t)}{d(x, \gamma_t)} \geq \frac{f(x) - f(y)}{d(x, y)}.$$

Taking supremum over y finishes the first part of the proof. To show that $|\partial f|$ is lsc, let $x_n \in \mathcal{H}$ converge to $x \in \operatorname{dom} |\partial f|$ and let $y \in \mathcal{H} \setminus \{x\}$. Then we have

$$\liminf_{n \to \infty} |\partial f(x_n)| \geq \liminf_{n \to \infty} \frac{\max\{f(x_n) - f(y), 0\}}{d(x_n, y)} \geq \frac{\max\{f(x) - f(y), 0\}}{d(x, y)}.$$

Now it suffices to take the supremum over y and we are done, by the first part of the lemma. □

Observe that $x \in \mathcal{H}$ is a minimizer of f if and only if $|\partial f|(x) = 0$. By the definition, $\operatorname{dom} |\partial f| \subset \operatorname{dom} f$ and it is easy to find a function f such that $\operatorname{dom} |\partial f| \neq \operatorname{dom} f$; see Exercise 5.1. On the other hand, the following lemma shows that the difference between these two domains cannot be too big. First recall the resolvent of f was defined in Definition 2.2.20 as

$$J_\lambda x := \operatorname*{arg\,min}_{y \in \mathcal{H}} \left[f(y) + \frac{1}{2\lambda} d(x, y)^2 \right], \qquad x \in \mathcal{H},$$

where $\lambda \in (0, \infty)$.

Lemma 5.1.3. *Let $f : \mathcal{H} \to (-\infty, \infty]$ be a convex lsc function. Then for every $x \in \mathcal{H}$ and $\lambda \in (0, \infty)$ we have $J_\lambda x \in \operatorname{dom} |\partial f|$ and*

$$|\partial f| (J_\lambda x) \leq \frac{d(x, J_\lambda x)}{\lambda}.$$

In particular, we see that $\operatorname{dom} |\partial f|$ *is dense in* $\operatorname{dom} f$.

Proof. Let $y \in \mathcal{H}$, then

$$\begin{aligned} f(J_\lambda x) - f(y) &\leq \frac{1}{2\lambda} d(x, y)^2 - \frac{1}{2\lambda} d(x, J_\lambda x)^2 \\ &\leq \frac{1}{2\lambda} d(y, J_\lambda x) \left[d(y, x) + d(x, J_\lambda x) \right], \end{aligned}$$

and hence we obtain

$$\begin{aligned} \limsup_{y \to J_\lambda x} \frac{\max\{0, f(J_\lambda x) - f(y)\}}{d(y, J_\lambda x)} &\leq \limsup_{y \to J_\lambda x} \frac{1}{2\lambda} \left[d(y, x) + d(x, J_\lambda x) \right] \\ &= \frac{d(x, J_\lambda x)}{\lambda}. \end{aligned}$$

This finishes the proof. □

Lemma 5.1.4. *Let $f\colon \mathcal{H} \to (-\infty, \infty]$ be a convex lsc function. Then for every $x \in \mathcal{H}$ and $\lambda \in (0, \infty)$, we have*

$$\frac{f(x) - f_\lambda(x)}{\lambda} \leq \frac{1}{2}|\partial f|(x)^2.$$

Proof. By Lemma 5.1.2 we have

$$\begin{aligned}
\frac{f(x) - f_\lambda(x)}{\lambda} &= \frac{f(x) - f(J_\lambda x)}{\lambda} - \frac{d(x, J_\lambda x)^2}{2\lambda^2} \\
&\leq |\partial f|(x)\frac{d(x, J_\lambda x)}{\lambda} - \frac{d(x, J_\lambda x)^2}{2\lambda^2} \\
&\leq \frac{1}{2}|\partial f|(x)^2 + \frac{d(x, J_\lambda x)^2}{2\lambda^2} - \frac{d(x, J_\lambda x)^2}{2\lambda^2} \\
&= \frac{1}{2}|\partial f|(x)^2 .
\end{aligned}$$

□

Combining Lemma 5.1.3, Lemma 2.2.23 and Lemma 5.1.4, we arrive at a heartwarming inequality

$$|\partial f|(J_\lambda x)^2 \leq \frac{d(x, J_\lambda x)^2}{\lambda^2} \leq 2\frac{f(x) - f_\lambda(x)}{\lambda} \leq |\partial f|(x)^2 \tag{5.1.3}$$

for every $x \in \mathcal{H}$, which we record here for future references.

Existence of a gradient flow semigroup

At first we show the existence of a flow (Theorem 5.1.6) and next prove its semigroup property (Proposition 5.1.15). Recall some notation.

Notation 5.1.5. Given a mapping $F\colon \mathcal{H} \to \mathcal{H}$, denote

$$F^{(n)} := F \circ \cdots \circ F, \quad n \in \mathbb{N},$$

where F appears n-times in the composition on the right-hand side. As a convention, we put $F^{(0)}x := x$ for every $x \in \mathcal{H}$.

Theorem 5.1.6 (Existence of a flow). *Let $(\mathcal{H}, d)$ be a Hadamard space and $f\colon \mathcal{H} \to (-\infty, \infty]$ be a convex lsc function. Assume $x_0 \in \overline{\operatorname{dom} f}$. Then the following limit exists*

$$S_t x_0 := \lim_{n\to\infty} J_{\frac{t}{n}}^{(n)} x_0, \quad t \in [0, \infty), \tag{5.1.4}$$

and defines a nonexpansive mapping $S_t\colon \overline{\operatorname{dom} f} \to \overline{\operatorname{dom} f}$. The limit in (5.1.4) is uniform with respect to t on bounded subintervals of $[0, \infty)$ and, if $x_0 \in \operatorname{dom} f$, we have the error estimate

$$d\left(S_t x_0, J_{\frac{t}{n}}^{(n)} x_0\right) \leq \sqrt{\frac{t}{n}\left[f(x_0) - f_{\frac{t}{n}}(x_0)\right]} \leq \frac{t}{\sqrt{2n}}|\partial f|(x_0), \tag{5.1.5}$$

for every $t \in [0, \infty)$ and $n \in \mathbb{N}$.

Proof. Fix $\tau > 0$. Choose $N \in \mathbb{N}$ and put $\lambda := \lambda(N) := \tau/N$. This gives a partition $\mathcal{P}_\lambda := \{n\lambda\}_{n\in\mathbb{N}}$ of the time interval $[0,\infty)$ with time step $\lambda > 0$. Also put $I_n^\lambda := ((n-1)\lambda, n\lambda]$ for every $n \in \mathbb{N}$. Let us first assume $x_0 \in \operatorname{dom} f$. We divide the proof into several steps.

Step 1: Put $x_n^\lambda := J_\lambda^{(n)} x_0$ for every $n \in \mathbb{N}_0$. Define the fractional part function

$$l_\lambda(t) := \frac{t-(n-1)\lambda}{\lambda}, \quad t \in I_n^\lambda,\ n \in \mathbb{N},$$

and, for each $y \in \mathcal{H}$, interpolate the distance function squared by putting

$$D_\lambda(t;y)^2 := (1-l_\lambda(t))\, d\left(y, x_{n-1}^\lambda\right)^2 + l_\lambda(t) d\left(y, x_n^\lambda\right)^2, \quad t \in I_n^\lambda,$$

and in the same way interpolate the function f, that is, put

$$\varphi_\lambda(t) := (1-l_\lambda(t))\, f\left(x_{n-1}^\lambda\right) + l_\lambda(t) f\left(x_n^\lambda\right), \quad t \in I_n^\lambda,$$

for every $n \in \mathbb{N}$.

Step 2: Define the function

$$\begin{aligned} r_\lambda(t) := {} & 2\,(1-l_\lambda(t))\left[f\left(x_{n-1}^\lambda\right) - f\left(x_n^\lambda\right) - \frac{1}{\lambda} d\left(x_n^\lambda, x_{n-1}^\lambda\right)^2\right] \\ & + (1-2l_\lambda(t))\,\frac{1}{\lambda} d\left(x_n^\lambda, x_{n-1}^\lambda\right)^2, \end{aligned}$$

and observe that

$$\int_{(n-1)\lambda}^{n\lambda} r_\lambda(t)\ \mathrm{d}t = \lambda\left[f\left(x_{n-1}^\lambda\right) - f\left(x_n^\lambda\right) - \frac{1}{\lambda} d\left(x_n^\lambda, x_{n-1}^\lambda\right)^2\right].$$

Consequently,

$$\int_0^\tau r_\lambda(t)\ \mathrm{d}t = \lambda \sum_{n=1}^N \left[f\left(x_{n-1}^\lambda\right) - f\left(x_n^\lambda\right) - \frac{1}{\lambda} d\left(x_n^\lambda, x_{n-1}^\lambda\right)^2\right].$$

Step 3: From (5.1.3) we get

$$\begin{aligned} f\left(x_{n-1}^\lambda\right) - f\left(x_n^\lambda\right) - \frac{1}{\lambda} d\left(x_n^\lambda, x_{n-1}^\lambda\right)^2 &= f\left(x_{n-1}^\lambda\right) - f_\lambda\left(x_{n-1}^\lambda\right) - \frac{1}{2\lambda} d\left(x_n^\lambda, x_{n-1}^\lambda\right)^2 \\ &\le \left[f\left(x_{n-1}^\lambda\right) - f_\lambda\left(x_{n-1}^\lambda\right)\right] - \left[f\left(x_n^\lambda\right) - f_\lambda\left(x_n^\lambda\right)\right], \end{aligned}$$

which yields

$$\begin{aligned} \sum_{n=1}^N \left[f\left(x_{n-1}^\lambda\right) - f\left(x_n^\lambda\right) - \frac{1}{\lambda} d\left(x_n^\lambda, x_{n-1}^\lambda\right)^2\right] &\le \left[f\left(x_0^\lambda\right) - f_\lambda\left(x_0^\lambda\right)\right] - \left[f\left(x_N^\lambda\right) - f_\lambda\left(x_N^\lambda\right)\right] \\ &\le f\left(x_0^\lambda\right) - f_\lambda\left(x_0^\lambda\right). \end{aligned}$$

This implies, again by (5.1.3), that

$$\int_0^\tau r_\lambda(t)\,\mathrm{d}t \leq \lambda\left[f\left(x_0^\lambda\right) - f_\lambda\left(x_0^\lambda\right)\right] \leq \frac{1}{2}\lambda^2 |\partial f|\left(x_0\right)^2 .$$

Step 4: We will now show that the functions D_λ and φ_λ defined above satisfy

$$\frac{1}{2}\frac{\mathrm{d}}{\mathrm{d}t} D_\lambda(t;y)^2 + \varphi_\lambda(t) - f(y) \leq \frac{1}{2} r_\lambda(t)\,, \tag{5.1.6}$$

for every $y \in \operatorname{dom} f$ and $t \in (0,\infty) \setminus \mathcal{P}_\lambda$. To prove this inequality, let us assume $t \in ((n-1)\lambda, n\lambda)$. Applying Lemma 2.2.23 yields

$$\begin{aligned}
&\frac{1}{2}\frac{\mathrm{d}}{\mathrm{d}t} D_\lambda(t;y)^2 + \varphi_\lambda(t) - f(y) \\
&\quad = \frac{1}{2\lambda}\left[d\left(x_n^\lambda, y\right)^2 - d\left(x_{n-1}^\lambda, y\right)^2\right] + \varphi_\lambda(t) - f(y) \\
&\quad \leq -\frac{1}{2\lambda} d\left(x_{n-1}^\lambda, x_n^\lambda\right)^2 + f(y) - f\left(x_n^\lambda\right) + \varphi_\lambda(t) - f(y) \\
&\quad = -\frac{1}{2\lambda} d\left(x_{n-1}^\lambda, x_n^\lambda\right)^2 + \left(1 - l_\lambda(t)\right)\left[f\left(x_{n-1}^\lambda\right) - f\left(x_n^\lambda\right)\right]
\end{aligned}$$

which implies (5.1.6).

Step 5: Let us now choose $M \in \mathbb{N}$ and put $\eta = \eta(M) := \tau/M$. We hence have another partition $\mathcal{P}_\eta := \{m\eta\}_{m\in\mathbb{N}}$ of the time interval $[0,\infty)$ with time step η. Put $x_m^\eta := J_\eta^{(m)} x_0$ for $m \in \mathbb{N}_0$. We will now interpolate the discrete time flows $(x_n^\lambda)_{n\in\mathbb{N}}$ and $(x_m^\eta)_{m\in\mathbb{N}}$ by setting

$$\underline{x}^\lambda(t) := x_{n-1}^\lambda, \quad \overline{x}^\lambda(t) := x_n^\lambda, \qquad t \in I_n^\lambda,$$

and likewise,

$$\underline{x}^\eta(t) := x_{m-1}^\eta, \quad \overline{x}^\eta(t) := x_m^\eta, \quad t \in I_m^\eta,$$

for every $n, m \in \mathbb{N}$.

Step 6: We *claim* that the function $D_{\lambda\eta}$ defined as

$$D_{\lambda\eta}(t,s)^2 := \left(1 - l_\eta(s)\right) D_\lambda\left(t; \underline{x}^\eta(s)\right)^2 + l_\eta(s) D_\lambda\left(t; \overline{x}^\eta(s)\right)^2, \quad t, s \in [0,\infty),$$

satisfies

$$\frac{\mathrm{d}}{\mathrm{d}t} D_{\lambda\eta}(t,t)^2 \leq r_\lambda(t) + r_\eta(t), \quad \text{for all } t \in (0,\infty) \setminus \mathcal{P}_\lambda \cup \mathcal{P}_\eta,$$

and therefore also

$$D_{\lambda\eta}(\tau,\tau)^2 \leq \int_0^\tau \left[r_\lambda(t) + r_\eta(t)\right] \mathrm{d}t\,. \tag{5.1.7}$$

Indeed, recall that

$$\varphi_\eta(s) = \left(1 - l_\eta(s)\right) f\left(\underline{x}^\eta(s)\right) + l_\eta(s) f\left(\overline{x}^\eta(s)\right),$$

and apply (5.1.6) first with $y := \underline{x}^\eta(s)$, and then with $y := \overline{x}^\eta(s)$, to get

$$\frac{1}{2}\frac{\mathrm{d}}{\mathrm{d}t} D_\lambda\left(t; \underline{x}^\eta(s)\right) + \varphi_\lambda(t) - f\left(\underline{x}^\eta(s)\right) \le \frac{1}{2} r_\lambda(t),$$

and,

$$\frac{1}{2}\frac{\mathrm{d}}{\mathrm{d}t} D_\lambda\left(t; \overline{x}^\eta(s)\right) + \varphi_\lambda(t) - f\left(\overline{x}^\eta(s)\right) \le \frac{1}{2} r_\lambda(t),$$

respectively, for every $s \in (0, \infty)$ and $t \in (0, \infty) \setminus \mathcal{P}_\lambda$. Taking the convex combination of the two last inequalities with coefficients $1 - l_\eta(s)$ and $l_\eta(s)$ gives

$$\frac{1}{2}\frac{\partial}{\partial t} D_{\lambda\eta}(t, s)^2 + \varphi_\lambda(t) - \varphi_\eta(s) \le \frac{1}{2} r_\lambda(t),$$

for every $s \in (0, \infty)$ and $t \in (0, \infty) \setminus \mathcal{P}_\lambda$. In the same way, applying (5.1.6) to $D_\eta(s; y)$ with $y := \underline{x}^\lambda(s)$, and then with $y := \overline{x}^\lambda(s)$, yields

$$\frac{1}{2}\frac{\partial}{\partial s} D_{\eta\lambda}(s, t)^2 + \varphi_\eta(s) - \varphi_\lambda(t) \le \frac{1}{2} r_\eta(s),$$

for every $t \in (0, \infty)$ and $s \in (0, \infty) \setminus \mathcal{P}_\eta$. Altogether we obtain

$$\frac{\partial}{\partial t} D_{\lambda\eta}(t, s)^2 + \frac{\partial}{\partial s} D_{\eta\lambda}(s, t)^2 \le r_\lambda(t) + r_\eta(s),$$

for every $t, s \in (0, \infty) \setminus \mathcal{P}_\lambda \cup \mathcal{P}_\eta$. This immediately implies the desired inequality and hence proves the *claim*.

Step 7: By (5.1.7) we have that $(\overline{x}^\lambda(\tau))_\lambda$ is Cauchy. Passing to the limit $\eta \to 0$ in (5.1.7) yields (5.1.5) and hence proves the whole theorem if $x_0 \in \operatorname{dom} f$. If $x_0 \in \overline{\operatorname{dom} f}$, we use a simple approximation argument. The proof is now complete. □

The error estimate (5.1.5) yields with the help of Proposition 2.2.26 that $t \mapsto S_t x_0$ is a continuous map from $[0, \infty)$ to $\mathcal{H}$.

Notation 5.1.7. We will often write $x_t := S_t x_0$ to denote a gradient flow curve. However, a word of caution is in order. At some places, the symbol x_t stands for a point on a geodesic $x_t := (1 - t)x_0 + t x_1$. But it is always clear from the context which case we have in mind.

Proposition 5.1.8 (Semigroup property of a flow). *Let $(\mathcal{H}, d)$ be a Hadamard space and $f\colon \mathcal{H} \to (-\infty, \infty]$ be a convex lsc function. Then (S_t) is a strongly continuous semigroup of nonexpansive mappings, that is,*
(i) $\lim_{t\to 0+} S_t x = x$,
(ii) $S_t(S_s x) = S_{s+t}x$, *for every* $s, t \in [0,\infty)$,
(iii) $d(S_t x, S_t y) \le d(x,y)$, *for each* $t \in [0,\infty)$,
for every $x, y \in \overline{\operatorname{dom} f}$.

Proof. (i) Follows immediately by Proposition 2.2.26 and (5.1.5). As for (ii), the proof is the same as in Theorem 4.3.3, only replace R_λ by J_λ and T_t by S_t. The nonexpansive property in (iii) is a direct consequence of Theorem 2.2.22 and (5.1.5). □

As a first and easy consequence of the semigroup property, we get that the function $t \mapsto f(x_t)$ is nonincreasing.

Corollary 5.1.9. *Let $f\colon \mathcal{H} \to (-\infty,\infty]$ be convex lsc and x_t be the gradient flow starting at a point $x_0 \in \overline{\operatorname{dom} f}$. Then the function $t \mapsto f(x_t)$ is nonincreasing. Indeed, since we have*

$$f\left(J_{\frac{t}{n}}^{(n)} x_0\right) \le f\left(J_{\frac{t}{n}} x_0\right) \le f\left(x_0\right),$$

and f is lsc, we obtain that $f(x_t) \le f(x_0)$. Monotonicity then follows by the semigroup property.

Properties of a gradient flow

Having constructed the gradient flow semigroup, we will proceed by showing its features. You may notice that many properties of the solution to the problem (5.0.2) in a Hilbert space transfer to our nonlinear setting.

Proposition 5.1.10 (Regularity of a flow). *Let $f\colon \mathcal{H} \to (-\infty,\infty]$ be a convex lsc function and $x_0 \in \overline{\operatorname{dom} f}$. Then the mapping $t \mapsto S_t x_0$ is locally Lipschitz on $(0,\infty)$ and Lipschitz on $[t_0,\infty)$ where t_0 is an arbitrary positive time.*

Proof. Let $x, y \in \mathcal{H}$. Applying iteratively Lemma 2.2.23 yields

$$d\left(J_{\frac{\varepsilon}{n}} x, y\right)^2 \le 2\frac{\varepsilon}{n}\left[f(y) - f\left(J_{\frac{\varepsilon}{n}} x\right)\right] + d(x,y)^2,$$
$$d\left(J_{\frac{\varepsilon}{n}}^{(2)} x, y\right)^2 \le 2\frac{\varepsilon}{n}\left[f(y) - f\left(J_{\frac{\varepsilon}{n}}^{(2)} x\right)\right] + d\left(J_{\frac{\varepsilon}{n}} x, y\right)^2,$$
$$\vdots$$
$$d\left(J_{\frac{\varepsilon}{n}}^{(n)} x, y\right)^2 \le 2\frac{\varepsilon}{n}\left[f(y) - f\left(J_{\frac{\varepsilon}{n}}^{(n)} x\right)\right] + d\left(J_{\frac{\varepsilon}{n}}^{(n-1)} x, y\right)^2.$$

Summing up these inequalities, dividing by ε^2 and putting $x := x_t$ and $y := x_t$ gives

$$\frac{d\left(J_{\frac{\varepsilon}{n}}^{(n)} x_t, x_t\right)^2}{\varepsilon^2} \leq 2 \frac{f(x_t) - f\left(J_{\frac{\varepsilon}{n}}^{(n)} x_t\right)}{\varepsilon},$$

and after taking $\limsup_{n\to\infty}$ we obtain

$$\frac{d(x_{t+\varepsilon}, x_t)^2}{\varepsilon^2} \leq 2 \frac{f(x_t) - f(x_{t+\varepsilon})}{\varepsilon}.$$

Since $t \mapsto f(x_t)$ is monotone by Corollary 5.1.9, it is differentiable almost everywhere on $[0, \infty)$ and thus

$$\limsup_{\varepsilon \to 0+} \frac{d(x_{t+\varepsilon}, x_t)}{\varepsilon} < \infty,$$

almost everywhere on $[0, \infty)$. Fix $s \in [0, \infty)$ and $C > 0$ such that

$$d(x_{s+\varepsilon}, x_s) \leq C\varepsilon,$$

for every sufficiently small $\varepsilon > 0$. Then, by the semigroup property,

$$d(x_{t+\varepsilon}, x_t) = d(S_{t-s} x_{s+\varepsilon}, S_{t-s} x_s) \leq d(x_{s+\varepsilon}, x_s) \leq C\varepsilon,$$

for every $t \in [s, \infty)$. In conclusion, the mapping $t \mapsto S_t x$ is locally Lipschitz on $(0, \infty)$ and Lipschitz on $[t_0, \infty)$ where t_0 is an arbitrary positive time. □

It is a well known fact in Hilbert spaces, that the evolution variational inequality completely characterizes the gradient flow. Next we see that this extends to Hadamard spaces.

Theorem 5.1.11 (Evolution variational inequality). *Let $(\mathcal{H}, d)$ be a Hadamard space and $f\colon \mathcal{H} \to (-\infty, \infty]$ be a convex lsc function. Assume $x_0 \in \overline{\operatorname{dom} f}$ and denote $x_t := S_t x_0$ for $t \in (0, \infty)$. Then $t \mapsto x_t$ is absolutely continuous on $(0, \infty)$ and satisfies*

$$\frac{1}{2} \frac{\mathrm{d}}{\mathrm{d}t} d(y, x_t)^2 + f(x_t) \leq f(y), \tag{5.1.9}$$

for almost every $t \in (0, \infty)$ and every $y \in \operatorname{dom} f$. Conversely, if an absolutely continuous curve $z\colon (0, \infty) \to \mathcal{H}$ with $\lim_{t\to 0+} z(t) = x_0$ satisfies (5.1.9), then $z_t = S_t x_0$ for every $t \in (0, \infty)$.

Proof. By an approximation argument, we can assume $x_0 \in \operatorname{dom} f$. Absolute continuity of $t \mapsto x_t$ follows by the local Lipschitz property. Choose $t_1 \in [0, \infty)$ and $t_2 \in [t_1, \infty)$. Integrating (5.1.6) from t_1 to t_2 and sending $\lambda \to 0$ gives

$$\frac{1}{2} d\left(y, x_{t_2}\right)^2 - \frac{1}{2} d\left(y, x_{t_1}\right)^2 + \int_{t_1}^{t_2} f(x_t)\,\mathrm{d}t \leq (t_2 - t_1) f(y), \tag{5.1.10}$$

for each $y \in \operatorname{dom} f$. Differentiating the above inequality with respect to t_2 yields (5.1.9). It remains to show that this solution is unique. Let $v\colon (0,\infty) \to \mathcal{H}$ be a absolutely continuous solution to (5.1.9) and integrate from $t-h$ to t to obtain

$$\frac{1}{2}d\left(y,v_t\right)^2 - \frac{1}{2}d\left(y,v_{t-h}\right)^2 + \int_{t-h}^{t} f\left(v_s\right)\mathrm{d}s \leq hf(y),$$

for each $y \in \operatorname{dom} f$. Dividing by h and passing to the limit gives

$$\limsup_{h\to 0+}\left[\frac{1}{2h}d\left(y,v_t\right)^2 - \frac{1}{2h}d\left(y,v_{t-h}\right)^2\right] + f\left(v_t\right) \leq f(y), \tag{5.1.11}$$

for each $t \in (0,\infty)$. By the same argument we arrive at

$$\limsup_{h\to 0+}\left[\frac{1}{2h}d\left(y,v_{t+h}\right)^2 - \frac{1}{2h}d\left(y,v_t\right)^2\right] + f\left(v_t\right) \leq f(y), \tag{5.1.12}$$

for each $t \in (0,\infty)$. Let now $u\colon (0,\infty) \to \operatorname{dom} f$ and $w\colon (0,\infty) \to \operatorname{dom} f$ be absolutely continuous curves satisfying (5.1.9) and

$$\lim_{t\to 0} u_t = \lim_{t\to 0} w_t = u_0 \in \overline{\operatorname{dom} f}.$$

Put $v_t := u_t$ and $y := w_t$ in (5.1.11) and $v_t := w_t$ and $y := v_t$ (5.1.12) and sum the resulting inequalities up. Then Exercise 5.2 gives

$$\frac{\mathrm{d}}{\mathrm{d}t}d\left(u_t,w_t\right)^2 \leq 0,$$

for almost every $t \in (0,\infty)$, which shows the uniqueness of the solution and the proof is complete. □

If we put $y := x_{t_1}$ in (5.1.10), then we see that the mapping $t \mapsto S_t x_0$ is $\frac{1}{2}$-Hölder on $[0,\infty)$. The Hölder constant $\frac{1}{2}$ is optimal in the sense that for each $\beta \in (\frac{1}{2},1)$ there exists a convex lsc function $f\colon [0,1] \to \mathbb{R}$, for instance

$$f(x) := 1 - x^{\alpha}, \quad x \in [0,1], \tag{5.1.13}$$

where $\alpha \in (0,1)$ is a number satisfying $\frac{1}{2-\alpha} < \beta$, such that the gradient flow is not β-Hölder. Indeed, the solution to the equation

$$\dot{x}(t) = -\nabla f\left(x(t)\right), \quad t \in [0,1],$$

with the initial condition $x(0) := 0$ is $x(t) := (\alpha(2-\alpha)t)^{\frac{1}{2-\alpha}}$, where $t \in [0,1]$, which is not a β-Hölder mapping.

The following is a quantified version of Corollary 5.1.9.

Proposition 5.1.12. *Let $f\colon \mathcal{H} \to (-\infty, \infty]$ be a convex lsc function. Given $x_0 \in \overline{\operatorname{dom}} f$, put $x_t := S_t x_0$. Then*

$$f(x_t) \le f(y) + \frac{1}{2t} d(y, x_0)^2 \tag{5.1.14}$$

for every $y \in \mathcal{H}$. Consequently, $\lim_{t\to\infty} f(x_t) = \inf_{\mathcal{H}} f$.

Proof. To show (5.1.14) we need to integrate (5.1.9) from 0 to τ and use monotonicity of $f(x_t)$ to get

$$\tau f(x_\tau) \le \int_0^\tau f(x_t)\,\mathrm{d}t \le \tau f(y) + \frac{1}{2} d(y, x_0)^2 - \frac{1}{2} d(y, x_\tau)^2 . \tag{5.1.15}$$

Then (5.1.14) follows. □

We get to a remarkable point, where one can see that a gradient flow moves at each point in the direction of the steepest descent of the function.

Theorem 5.1.13. *Let $(\mathcal{H}, d)$ be a Hadamard space and $f\colon \mathcal{H} \to (-\infty, \infty]$ be a convex lsc function. Given $x_0 \in \overline{\operatorname{dom}} f$, put $x_t := S_t x_0$. Then,*

$$|\partial f|(x_t) = \lim_{h\to 0+} \frac{d(x_{t+h}, x_t)}{h}, \tag{5.1.16}$$

as well as,

$$|\partial f|(x_t) = \lim_{h\to 0+} \frac{f(x_t) - f(x_{t+h})}{d(x_{t+h}, x_t)}, \tag{5.1.17}$$

and also,

$$|\partial f|(x_t)^2 = \lim_{h\to 0+} \frac{f(x_t) - f(x_{t+h})}{h}, \tag{5.1.18}$$

for every $t \in (0, \infty)$.

Proof. Step 1: Integrating (5.1.9) from t to $t + h$ and dividing by h yields

$$\begin{aligned}\frac{1}{h}\int_0^h f(x_{t+s})\,\mathrm{d}s - f(y) &\le \frac{1}{2h}\left[d(x_t, y)^2 - d(x_{t+h}, y)^2\right] \\ &\le \left[d(x_t, y) + d(x_{t+h}, y)\right] \frac{d(x_{t+h}, x_t)}{2h},\end{aligned} \tag{5.1.19}$$

for every $y \in \operatorname{dom} f$. Taking $\liminf_{h\to 0+}$ and recalling that $t \mapsto f(x_t)$ is lsc, we have

$$f(x_t) - f(y) \le d(x_t, y) \liminf_{h\to 0+} \frac{d(x_{t+h}, x_t)}{h}, \tag{5.1.20}$$

for every $y \in \operatorname{dom} f$. Taking supremum over $y \in \operatorname{dom} f$ gives

$$|\partial f|(x_t) \leq \liminf_{h\to 0+} \frac{d(x_{t+h}, x_t)}{h},$$

by Lemma 5.1.2. On the other hand if we put $y := x_t$ in (5.1.19) and rescale the integrand, then

$$\begin{aligned}\frac{d(x_{t+h}, x_t)^2}{2h^2} &\leq \frac{1}{h}\int_0^1 [f(x_t) - f(x_{t+hs})]\,\mathrm{d}s \\ &\leq |\partial f|(x_t) \int_0^1 \frac{d(x_{t+hs}, x_t)}{hs} s\,\mathrm{d}s,\end{aligned}$$

where we again used Lemma 5.1.2. Taking $\limsup_{h\to 0+}$ yields

$$\begin{aligned}\limsup_{h\to 0+} \frac{d(x_{t+h}, x_t)^2}{2h^2} &\leq |\partial f|(x_t) \int_0^1 \limsup_{h\to 0+} \frac{d(x_{t+hs}, x_t)}{hs} s\,\mathrm{d}s \\ &\leq |\partial f|(x_t) \int_0^1 \limsup_{h'\to 0+} \frac{d(x_{t+h'}, x_t)}{h'} s\,\mathrm{d}s \\ &\leq \frac{1}{2}|\partial f|(x_t) \limsup_{h\to 0+} \frac{d(x_{t+h}, x_t)}{h}.\end{aligned}$$

We hence get

$$\limsup_{h\to 0+} \frac{d(x_{t+h}, x_t)}{h} \leq |\partial f|(x_t).$$

The one-sided limit

$$\lim_{h\to 0+} \frac{d(x_{t+h}, x_t)}{h}$$

therefore exists and is equal to $|\partial f|(x_t)$, which proves (5.1.16).

Step 2: We now prove (5.1.17). The inequality

$$\limsup_{h\to 0+} \frac{f(x_t) - f(x_{t+h})}{d(x_t, x_{t+h})} \leq |\partial f|(x_t)$$

follows immediately from the definition of the slope. To show the converse, choose $x \in \operatorname{dom} f$ such that $f(x_t) > f(x)$ and choose $\varepsilon_1 > 0$ such that $3\varepsilon_1 < d(x, x_t)$. Then consider $h \in (0, \infty)$ such that

$$d\left(x_t, J^{(n)}_{\frac{h}{n}} x_t\right) < \varepsilon_1,$$

for every $n \in \mathbb{N}$, which for a fixed $n \in \mathbb{N}$ and $k = 1, \ldots, n$ implies that

$$\lambda_k := \frac{d\left(J_{\frac{h}{n}}^{(k)} x_t, J_{\frac{h}{n}}^{(k-1)} x_t\right)}{d\left(x, J_{\frac{h}{n}}^{(k-1)} x_t\right)},$$

which satisfies $\lambda_k \in [0, 1)$. Denote

$$y_k := (1 - \lambda_k)\, J_{\frac{h}{n}}^{(k-1)} x_t + \lambda_k x,$$

and observe

$$d\left(y_k, J_{\frac{h}{n}}^{(k-1)} x_t\right) = d\left(J_{\frac{h}{n}}^{(k)} x_t, J_{\frac{h}{n}}^{(k-1)} x_t\right),$$

which further gives $f(J_{\frac{h}{n}}^{(k)} x_t) \leq f(y_k)$, for $k = 1, \ldots, n$. Since f is convex, we have

$$f\left(J_{\frac{h}{n}}^{(k)} x_t\right) \leq (1 - \lambda_k)\, f\left(J_{\frac{h}{n}}^{(k-1)} x_t\right) + \lambda_k f(x),$$

and, after the nth iteration, we arrive at

$$f\left(J_{\frac{h}{n}}^{(n)} x_t\right) \leq \left[\prod_{k=1}^{n} (1 - \lambda_k)\right] f(x_t) + \left[1 - \prod_{k=1}^{n} (1 - \lambda_k)\right] f(x).$$

We already know that $d(J_{\frac{h}{n}}^{(k-1)} x_t, x) \leq d(x_t, x) + \varepsilon_1$ and compute

$$\begin{aligned}
\log\left[\prod_{k=1}^{n} (1 - \lambda_k)\right] &= \sum_{k=1}^{n} \log(1 - \lambda_k) \\
&\leq -\sum_{k=1}^{n} \lambda_k \\
&\leq -\sum_{k=1}^{n} \frac{d\left(J_{\frac{h}{n}}^{(k)} x_t, J_{\frac{h}{n}}^{(k-1)} x_t\right)}{d(x_t, x) + \varepsilon_1} \\
&\leq -\frac{d\left(J_{\frac{h}{n}}^{(n)} x_t, x_t\right)}{d(x_t, x) + \varepsilon_1}.
\end{aligned}$$

Then the previous inequality reads

$$f\left(J_{\frac{h}{n}}^{(n)} x_t\right) \leq f(x) + \exp\left(-\frac{d\left(J_{\frac{h}{n}}^{(n)} x_t, x_t\right)}{d(x_t, x) + \varepsilon_1}\right) [f(x_t) - f(x)],$$

and after letting $n \to \infty$,

$$f(x_{t+h}) \leq f(x) + \exp\left(-\frac{d(x_{t+h}, x_t)}{d(x_t, x) + \varepsilon_1}\right) [f(x_t) - f(x)].$$

Observe that if $\varepsilon_2 \in (0,1)$, then $\exp(-\alpha) \le 1-(1-\varepsilon_2)\alpha$ for sufficiently small $\alpha \in [0,1)$. So we can further decrease h to have

$$\frac{d(x_{t+h}, x_t)}{d(x_t, x) + \varepsilon_1}$$

sufficiently small and we then have

$$(1-\varepsilon_2)\frac{f(x_t) - f(x)}{d(x_t, x) + \varepsilon_1} \le \frac{f(x_t) - f(x_{t+h})}{d(x_t, x_{t+h})}.$$

Letting $h \to 0+$ gives

$$(1-\varepsilon_2)\frac{f(x_t) - f(x)}{d(x_t, x) + \varepsilon_1} \le \liminf_{h\to 0+} \frac{f(x_t) - f(x_{t+h})}{d(x_t, x_{t+h})}.$$

Now send ε_1 and ε_2 to 0 and take $\limsup_{x\to x_t}$ to obtain

$$|\partial f|(x_t) = \limsup_{x\to x_t} \frac{f(x_t) - f(x)}{d(x_t, x)} \le \liminf_{h\to 0+} \frac{f(x_t) - f(x_{t+h})}{d(x_t, x_{t+h})},$$

which finishes the proof of (5.1.17).

Step 3: We easily get (5.1.18) from (5.1.16) and (5.1.17) since

$$\frac{f(x_t) - f(x_{t+h})}{h} = \frac{f(x_t) - f(x_{t+h})}{d(x_t, x_{t+h})} \cdot \frac{d(x_t, x_{t+h})}{h}.$$

The proof is complete. □

A slope has a similar decay as $f(x_t)$ in Proposition 5.1.12.

Proposition 5.1.14. *Let $f\colon \mathcal{H} \to (-\infty, \infty]$ be a convex lsc function and $x_0 \in \overline{\operatorname{dom} f}$. Put $x_t := S_t x_0$. Then the function $t \mapsto |\partial f|(x_t)$ is nonincreasing and*

$$|\partial f|(x_t)^2 \le |\partial f|(y)^2 + \frac{1}{t^2} d(y, x_0)^2,$$

for every $y \in \mathcal{H}$.

Proof. By (5.1.3) we have

$$|\partial f|\left(J_{\frac{t}{n}}^{(n)} x\right) \le |\partial f|\left(J_{\frac{t}{n}} x\right) \le |\partial f|(x),$$

and since $|\partial f|$ is lsc (Lemma 5.1.2), we obtain $|\partial f|(x_t) \le |\partial f|(x)$. Monotonicity then follows by the semigroup property.

Multiply (5.1.18) by t and integrate from 0 to τ and use the monotonicity of $|\partial f|$ to get

$$\frac{\tau^2}{2}|\partial f|(x_\tau)^2 \le \int_0^\tau t\,|\partial f|(x_t)^2\,\mathrm{d}t$$
$$\le -\int_0^\tau t\,[f(x_t)]'\,\mathrm{d}t$$
$$= \int_0^\tau f(x_t)\,\mathrm{d}t - \tau f(x_\tau)\,,$$

and by (5.1.15) we have, for each $y \in \mathcal{H}$,

$$\le \tau f(y) + \frac{1}{2}d(y,x_0)^2 - \tau f(x_\tau) - \frac{1}{2}d(y,x_\tau)^2\,,$$
$$\le \tau|\partial f|(y)d(y,x_\tau) - \frac{1}{2}d(y,x_\tau)^2 + \frac{1}{2}d(y,x_0)^2$$
$$\le \frac{\tau^2}{2}|\partial f|(y)^2 + \frac{1}{2}d(y,x_0)^2\,.$$

The proof is now complete. □

It is obvious that $x_t \in \operatorname{dom}|\partial f|$ for every $t \in (0,\infty)$. By the Lipschitz property of $t \mapsto x_t$ and (5.1.20) with $y := x_{t+h}$ we get that the function $t \mapsto f(x_t)$ is Lipschitz on (t_0,∞) for an arbitrary $t_0 > 0$. Furthermore, it is convex by (5.1.18).

In the end, we take a look at gradient flows of *strongly* convex lsc functions, which we have already met in Exercise 2.6.

Proposition 5.1.15. *Let $f\colon \mathcal{H} \to (-\infty,\infty]$ be a strongly convex function with parameter κ. Then the corresponding gradient flow satisfies*

$$d(S_t x, S_t y) \le \exp(-t\kappa)d(x,y)\,, \tag{5.1.21}$$

for every $x, y \in \overline{\operatorname{dom}} f$, and also a stronger version of the evolution variational inequality

$$\frac{1}{2}\frac{\mathrm{d}}{\mathrm{d}t}d(y,x_t)^2 + \frac{\kappa}{2}d(y,x_t)^2 + f(x_t) \le f(y)\,,$$

for almost every $t \in (0,\infty)$ and every $y \in \operatorname{dom} f$.

Proof. See Exercise 5.4. □

We know by Proposition 2.2.17 that a lsc strongly convex function $f\colon \mathcal{H} \to (-\infty,\infty]$ has a unique minimizer $z \in \mathcal{H}$. By (5.1.21) we have the following error estimate for the flow

$$d(S_t x, z) \le \exp(-t\kappa)d(x,z)\,, \tag{5.1.22}$$

for every $x \in \overline{\operatorname{dom} f}$. Here again $\kappa \in (0, \infty)$ is the parameter of the strong convexity of f.

Large time behavior of a gradient flow

Next we study what happens with a flow if $t \to \infty$. It is natural to expect that it converges to a minimizer of f provided f attains its minimum. It is indeed the case, but unfortunately, the flow $(S_t x_0)$ convergences only weakly to a minimizer; see Definition 3.1.9. Compare with the convergence of $(J_\lambda x_0)$ as $\lambda \to \infty$ given in Theorem 2.2.25.

Theorem 5.1.16 (Gradient flow convergence). *Let $(\mathcal{H}, d)$ be a Hadamard space and $f : \mathcal{H} \to (-\infty, \infty]$ be a convex lsc function attaining its minimum. Then, given a starting point $x_0 \in \overline{\operatorname{dom} f}$, the gradient flow $x_t := S_t x_0$ weakly converges to a minimizer of f as $t \to \infty$.*

Proof. Choose a sequence $(t_n) \subset [0, \infty)$ such that $t_n \to \infty$. The set of minimizers $\operatorname{Min} f$ is nonempty by the assumptions and $(x_{t_n})_n$ is Fejér monotone with respect to $\operatorname{Min} f$ since S_t is nonexpansive. Furthermore, the sequence $(x_{t_n})_n$ is minimizing, that is,

$$\lim_{n \to \infty} f\left(x_{t_n}\right) = \inf_{y \in \mathcal{H}} f(y),$$

by Proposition 5.1.14. Since f is weakly lsc (Lemma 3.2.3), it follows that all weak cluster points of (x_{t_n}) lie in $\operatorname{Min} f$. Apply Proposition 3.2.6 (iii) to get that (x_{t_n}) weakly converges to a point $z \in \operatorname{Min} f$.

If we now choose another sequence $(t_n') \subset [0, \infty)$ such that $t_n' \to \infty$, we have to show that $x_{t_n'} \xrightarrow{w} z$. Since the sequence $(x_{t_n}) \cup (x_{t_n'})$ is also Fejér monotone with respect to $\operatorname{Min} f$, we obtain that it weakly converges to z using the same argument as above. Consequently, $x_{t_n'} \xrightarrow{w} z$. The proof is complete. □

There exists a convex lsc function $f : \ell_2 \to (-\infty, \infty]$ attaining its minimum and $x_0 \in \overline{\operatorname{dom} f}$ such that the corresponding gradient flow semigroup converges weakly to a minimizer, but fails to converge strongly [23]. If the function in question is strongly convex on (bounded subsets of) its domain, then the gradient flow converges strongly by (5.1.22). The reader is invited to give an elementary proof of this fact in Exercise 5.5. We also obtain strong convergence provided the underlying space be locally compact.

Note that the proximal point algorithm is a discrete version of the gradient flow (see Section 6.3) and it is not surprising then that it has the same asymptotic behavior.

In the end, we give an application of gradient flows in Hadamard spaces into Kähler geometry. For the details, the reader is referred to [190, 191].

Example 5.1.17 (Calabi flow). Let (M^{2n}, ω) be a compact Kähler manifold. The set of *Kähler potentials*

$$\mathcal{K} := \left\{ \varphi \in \mathcal{C}^\infty(M, \mathbb{R}) \colon \omega_\varphi := \omega + \sqrt{-1}\partial\bar{\partial}\varphi > 0 \right\}$$

is equipped with the L^2-metric as follows. Given $\varphi \in \mathcal{K}$, we can identify the tangent space $T_\varphi\mathcal{K}$ with $\mathcal{C}^\infty(M, \mathbb{R})$ and put

$$\langle \psi_1, \psi_2 \rangle := \int_M \psi_1 \psi_2 \, \omega_\varphi^n ,$$

which gives a Riemannian metric on $\mathcal{K}$; see [70, 135, 183]. Since $\mathcal{K}$ is not complete with respect to this metric, we consider its completion $\overline{\mathcal{K}}$, which turns out to be a Hadamard space.

Denote now $s(\omega_\varphi)$ the scalar curvature of a metric ω_φ and put

$$\bar{s} := \frac{1}{\operatorname{Vol}(M)} \int_M s\left(\omega_\varphi\right) \omega_\varphi^n ,$$

which is independent of φ. A solution $\varphi \colon [0, \infty) \to \mathcal{K}$ to the *Calabi equation*

$$\frac{\partial}{\partial t}\varphi(t) = s\left(\omega_{\varphi(t)}\right) - \bar{s} ,$$

is called a *Calabi flow* and it can be equivalently defined as the gradient flow of the *Mabuchi functional*

$$f(\varphi) := -\int_0^1 \int_M \frac{\partial \varphi(t)}{\partial t} \left[s\left(\omega_{\varphi(t)}\right) - \bar{s} \right] \omega_{\varphi(t)}^n \wedge \mathrm{d}t ,$$

where $\varphi(\cdot) \colon [0, 1] \to \mathcal{K}$ is a curve with $\varphi(0) = 0$ and $\varphi(1) = \varphi$. The Mabuchi functional $f \colon \mathcal{K} \to \mathbb{R}$ is convex and lsc and we can extend it onto $\overline{\mathcal{K}}$ by

$$\bar{f}(\varphi) := \begin{cases} f(\varphi), & \text{whenever } \varphi \in \mathcal{K} \\ \liminf_{\mathcal{K} \ni \varphi_n \to \varphi} f\left(\varphi_n\right), & \text{elsewhere.} \end{cases}$$

Then $\bar{f} \colon \overline{\mathcal{K}} \to \mathbb{R}$ is also convex and lsc.

Recall that a conjecture of S. Donaldson [71] states that if (M^{2n}, ω) is a compact Kähler manifold of constant scalar curvature, then the Calabi flow with any initial condition exists for all times and converges to the constant scalar curvature metric. While this conjecture remains open, we can apply Theorem 5.1.16 to conclude that the gradient flow of the functional $\bar{f}$ with any initial condition exists for all times and weakly converges to a minimizer $\bar{\varphi} \in \overline{\mathcal{K}}$ of $\bar{f}$.

A next step towards the full Donaldson conjecture is to show the minimizer of the Mabuchi energy remains unique in the completion.

5.2 Mosco convergence and its consequences

In this section we consider a *sequence* of convex lsc functions $f^n \colon \mathcal{H} \to (-\infty, \infty]$, where $n \in \mathbb{N}$, instead of a single function. Let us first establish the notation.

Notation 5.2.1. Given a sequence of convex lsc functions (f^n), the resolvent of the function f^n, where $n \in \mathbb{N}$, will be denoted J_λ^n, and its semigroup S_t^n. We study a situation, when the sequence (f^n) converges (in some sense) to a limit function $f \colon \mathcal{H} \to (-\infty, \infty]$, whose resolvent and semigroup are denoted J_λ and S_t, respectively.

We establish a relationship among the sequences (f^n) and (J_λ^n) and (S_t^n). Namely, we show that the Mosco convergence of functions implies the convergence of the resolvents, and that the convergence of the resolvents implies the convergence of the semigroups.

Definition 5.2.2 (Γ-convergence). A sequence (f^n) of functions $f^n \colon \mathcal{H} \to (-\infty, \infty]$ is said to Γ-converge to a function $f \colon \mathcal{H} \to (-\infty, \infty]$ if, for each $x \in \mathcal{H}$, we have:
(Γ1) $f(x) \leq \liminf_{n\to\infty} f^n(x_n)$, whenever $x_n \to x$, and
(Γ2) there exists $(y_n) \subset \mathcal{H}$ such that $y_n \to x$ and $f^n(y_n) \to f(x)$.

Like in Hilbert spaces, Γ-convergence preserves convexity and the limit function is always lsc; see Exercises 5.6 and 5.7. We will however use a stronger type of convergence, called the Mosco convergence.

Definition 5.2.3 (Mosco convergence). The sequence (f^n) converges to f in the sense of *Mosco* if, for each $x \in \mathcal{H}$, we have:
(M1) $f(x) \leq \liminf_{n\to\infty} f^n(x_n)$, whenever $x_n \xrightarrow{w} x$, and
(M2) there exists $(y_n) \subset \mathcal{H}$ such that $y_n \to x$ and $f^n(y_n) \to f(x)$.

The advantage of the Mosco convergence is that it implies the convergence of the Moreau–Yosida envelopes and resolvents. As a matter of fact, these conditions are equivalent in Hilbert spaces [12, Theorem 3.26]. It is not known whether the same holds in Hadamard spaces; see Question 5.2.5 below.

Theorem 5.2.4. *Let $(\mathcal{H}, d)$ be a Hadamard space and $f^n \colon \mathcal{H} \to (-\infty, \infty]$ be convex lsc functions, for each $n \in \mathbb{N}$. If $f^n \to f$ in the sense of Mosco, as $n \to \infty$, then*

$$\lim_{n\to\infty} f_\lambda^n(x) = f_\lambda(x)\,, \tag{5.2.23}$$

and,

$$\lim_{n\to\infty} J_\lambda^n x = J_\lambda x\,, \tag{5.2.24}$$

for every $\lambda \in (0, \infty)$ and $x \in \mathcal{H}$. Here J_λ^n stands for the resolvent of f^n and J_λ for the resolvent of f.

Proof. Step 1: We will first show that the sequence $(J_\lambda^n x)_n$ is bounded. To do so, we need the following claim, which extends Lemma 2.2.13.

Claim: Given $x_0 \in \mathcal{H}$, there exist $\alpha, \beta \in (0, \infty)$ such that

$$f^n(y) \geq -\alpha d(y, x_0) - \beta\,, \tag{5.2.25}$$

for every $y \in \mathcal{H}$ and $n \in \mathbb{N}$. Indeed, assume that this is not the case, that is, for each $k \in \mathbb{N}$, there exist $n_k \in \mathbb{N}$ and $x_k \in \mathcal{H}$ such that

$$f^{n_k}(x_k) + k\left[d\left(x_k, x_0\right) + 1\right] < 0.$$

Without loss of generality we may assume $n_k \to \infty$ as $k \to \infty$, for otherwise a contradiction follows already from Lemma 2.2.13. If (x_k) were bounded, then there exist $\widehat{x} \in \mathcal{H}$ and a subsequence of (x_k), still denoted (x_k), such that $x_k \xrightarrow{w} \widehat{x}$. By the Mosco convergence of (f^n) we have

$$f\left(\widehat{x}\right) \leq \liminf_{k\to\infty} f^{n_k}\left(x_k\right) \leq -\limsup_{k\to\infty} k\left[d\left(x_k, x_0\right) + 1\right] \leq -\infty,$$

which is impossible. Assume therefore (x_k) is unbounded. Choose $y_0 \in \mathcal{H}$ and find $y_k \to y_0$ such that $f^{n_k}(y_k) \to f(y_0)$. Put

$$z_k := (1 - t_k)y_k + t_k x_k, \quad \text{with } t_k := \frac{1}{\sqrt{k}d(x_k, y_k)}.$$

Then $z_k \to y_0$. By convexity,

$$\begin{aligned} f^{n_k}(z_k) &\leq (1 - t_k)f^{n_k}(y_k) + t_k f^{n_k}(x_k) \\ &\leq (1 - t_k)f^{n_k}(y_k) - t_k k\left[d(x_k, x_0) + 1\right] \\ &\leq (1 - t_k)f^{n_k}(y_k) - \sqrt{k}\frac{d(x_k, x_0) + 1}{d(x_k, y_k)}\,. \end{aligned}$$

Hence,

$$f(y_0) \leq \liminf_{k\to\infty} f^{n_k}(z_k) \leq -\infty,$$

which is not possible, either. This proves the *claim.*

Step 2: For each $n \in \mathbb{N}$ we hence have

$$f^n\left(J_\lambda^n x\right) \geq -\alpha d(J_\lambda^n x, x_0) - \beta.$$

Choose a sequence $(u_n) \subset \mathcal{H}$ such that $u_n \to x_0$ and $f^n(u_n) \to f(x_0)$. From the definition of $J_\lambda^n x$, we have

$$f^n(u_n) + \frac{1}{2\lambda}d(x, u_n)^2 \geq f^n\left(J_\lambda^n x\right) + \frac{1}{2\lambda}d\left(x, J_\lambda^n x\right)^2,$$

and furthermore,

$$f^n(u_n) + \alpha d(J_\lambda^n x, x_0) + \beta + \frac{1}{2\lambda}d(x, u_n)^2 \geq \frac{1}{2\lambda}d\left(x, J_\lambda^n x\right)^2,$$

which implies that the sequence $(J_\lambda^n x)_n$ is bounded.

Let $c \in \mathcal{H}$ be a weak cluster point of $(J_\lambda^n x)_n$. Its existence is guaranteed by boundedness of the sequence. Since $f^n \to f$ in the sense of Mosco, there exists a sequence $(y_n) \subset \mathcal{H}$ such that $y_n \to J_\lambda x$ and $f^n(y_n) \to f(J_\lambda x)$. Then

$$\begin{aligned}
\limsup_{n\to\infty} f_\lambda^n(x) &\le \limsup_{n\to\infty} \left[f^n(y_n) + \frac{1}{2\lambda} d(x, y_n)^2 \right] \\
&= f(J_\lambda x) + \frac{1}{2\lambda} d(x, J_\lambda x)^2 \\
&\le f(c) + \frac{1}{2\lambda} d(x, c)^2 \\
&\le \liminf_{n\to\infty} \left[f^n(J_\lambda^n x) + \frac{1}{2\lambda} d(x, J_\lambda^n x)^2 \right], \qquad (5.2.26)
\end{aligned}$$

which gives $J_\lambda x = c$, by the uniqueness of $J_\lambda x$. Hence, since c was arbitrary, the whole sequence $(J_\lambda^n x)_n$ weakly converges to $J_\lambda x$. Furthermore,

$$\begin{aligned}
\limsup_{n\to\infty} \frac{1}{2\lambda} d(x, J_\lambda^n x)^2 &\le \limsup_{n\to\infty} (-f^n(J_\lambda^n x)) + \limsup_{n\to\infty} f^n(y_n) \\
&\quad + \limsup_{n\to\infty} \frac{1}{2\lambda} d(x, y_n)^2 \\
&\le -\liminf_{n\to\infty} f^n(J_\lambda^n x) + f(J_\lambda x) + \frac{1}{2\lambda} d(x, J_\lambda x)^2 \\
&\le \frac{1}{2\lambda} d(x, J_\lambda x)^2 \\
&\le \liminf_{n\to\infty} \frac{1}{2\lambda} d(x, J_\lambda^n x)^2 . \qquad (5.2.27)
\end{aligned}$$

Proposition 3.1.6 and (5.2.27) give together the strong convergence

$$J_\lambda^n x \to J_\lambda x, \quad \text{as } n \to \infty,$$

which proves (5.2.24). Finally, inequality (5.2.26) immediately gives (5.2.23) and the proof is complete. □

As we have already mentioned, each of (5.2.23) and (5.2.24) in Theorem 5.2.4 in turn implies the Mosco convergence of (f^n) as long as the underlying space is Hilbert. We raise the question whether the same is true in a general Hadamard space.

Question 5.2.5. Do (5.2.23) or (5.2.24) in Theorem 5.2.4 in turn imply $f^n \to f$ in the sense of Mosco?

Definition 5.2.6 (Mosco convergence of sets). A sequence of closed convex sets $C_n \subset \mathcal{H}$ is said to *Mosco converge* to a set $C \subset \mathcal{H}$ if the indicator functions ι_{C_n} converge in the sense of Mosco to the indicator function ι_C.

Example 5.2.7 (Monotone sequences of sets). Let (C_n) be a sequence of convex closed subsets of a Hadamard space. If (C_n) is nonincreasing, then it converges in the sense of Mosco to its intersection. Likewise, if (C_n) is nondecreasing, it converges in the sense of Mosco to the closure of its union. These facts are rather straightforward to prove; see [148, Lemma 1.2, Lemma 1.3] for the linear case.

Corollary 5.2.8. *Let $(\mathcal{H}, d)$ be a Hadamard space. Assume $C_n \subset \mathcal{H}$ are convex closed sets for each $n \in \mathbb{N}$. If the sequence (C_n) converges to a set $C \subset \mathcal{H}$ in the sense of Mosco, then the set C is closed convex and:*
(i) $d(x, C_n) \to d(x, C)$,
(ii) $P_{C_n} x \to P_C x$,
for each $x \in \mathcal{H}$.

Proof. The Mosco convergence of functions preserves convexity and the limit function is always lsc. This gives the first statement. The rest follows immediately from the fact that the Moreau–Yosida envelope (with $\lambda := \frac{1}{2}$) of the indicator function is the distance function squared and the resolvent of the indicator function is the nearest point mapping. □

The convergence in Corollary 5.2.8 (i) is called *Frolík–Wijsman*. Hence the Mosco convergence of convex closed sets implies the Frolík–Wijsman convergence.

We next show that the convergence of the resolvents implies the convergence of the semigroups.

Theorem 5.2.9. *Let $(\mathcal{H}, d)$ be a Hadamard space. Assume $f: \mathcal{H} \to (-\infty, \infty]$ and $f^n: \mathcal{H} \to (-\infty, \infty]$, for $n \in \mathbb{N}$, are convex lsc functions. We use Notation 5.2.1. Assume that for every $x \in \overline{\operatorname{dom} f}$ and $\lambda \in (0, \infty)$ we have*

$$\lim_{n\to\infty} J_\lambda^n x = J_\lambda x \,. \tag{5.2.28}$$

Then

$$\lim_{n\to\infty} S_t^n x = S_t x \,, \tag{5.2.29}$$

for every $x \in \overline{\operatorname{dom} f}$ and $t \in (0, \infty)$.

Proof. Step 1: Assume first $x \in \operatorname{dom} |\partial f|$. We then have

$$\begin{aligned} d\left(S_t^n x, S_t x\right) &\le d\left(S_t^n x, S_t^n \left(J_\lambda^n x\right)\right) + d\left(S_t^n \left(J_\lambda^n x\right), S_t x\right) \\ &\le d\left(x, J_\lambda^n x\right) + d\left(S_t^n \left(J_\lambda^n x\right), S_t x\right) \,, \end{aligned}$$

for every $\lambda \in (0, \infty)$ and $n \in \mathbb{N}$. The second term on the right-hand side can be further estimated

$$\begin{aligned} d\left(S_t^n \left(J_\lambda^n x\right), S_t x\right) &\le d\left(S_t^n \left(J_\lambda^n x\right), \left(J_{\frac{t}{k}}^n\right)^{(k)} J_\lambda^n x\right) + d\left(\left(J_{\frac{t}{k}}^n\right)^{(k)} J_\lambda^n x, \left(J_{\frac{t}{k}}^n\right)^{(k)} x\right) \\ &\quad + d\left(\left(J_{\frac{t}{k}}^n\right)^{(k)} x, \left(J_{\frac{t}{k}}\right)^{(k)} x\right) + d\left(\left(J_{\frac{t}{k}}\right)^{(k)} x, S_t x\right) \,. \end{aligned}$$

By (5.1.5) and (5.1.3),

$$d\left(S_t^n\left(J_\lambda^n x\right), \left(J_{\frac{t}{k}}^n\right)^{(k)} J_\lambda^n x\right) \le \frac{t}{\sqrt{2k}} \left|\partial f^n\right|\left(J_\lambda^n x\right) \le \frac{t}{\sqrt{2k}} \frac{d\left(J_\lambda^n x, x\right)}{\lambda}.$$

Furthermore, inequality (5.1.5) also reads

$$d\left(\left(J_{\frac{t}{k}}\right)^{(k)} x, S_t x\right) \le \frac{t}{\sqrt{2k}} \left|\partial f\right|(x).$$

Altogether we obtain,

$$\begin{aligned} d\left(S_t^n x, S_t x\right) \le & 2d\left(x, J_\lambda^n x\right) + \frac{t}{\sqrt{2k}} \frac{d\left(J_\lambda^n x, x\right)}{\lambda} \\ & + d\left(\left(J_{\frac{t}{k}}^n\right)^{(k)} x, \left(J_{\frac{t}{k}}\right)^{(k)} x\right) + \frac{t}{\sqrt{2k}} \left|\partial f\right|(x). \end{aligned}$$

Step 2: Now fix $\varepsilon > 0$ and choose $\lambda_0 \in (0,1)$ so that $\sqrt{\lambda_0}|\partial f|(x) < \varepsilon$. By the assumption (5.2.28) and by inequality (5.1.3), we have

$$\lim_{n\to\infty} \frac{d\left(x, J_\lambda^n x\right)}{\sqrt{\lambda}} = \frac{d\left(x, J_\lambda x\right)}{\sqrt{\lambda}} \le |\partial f|(x),$$

for each $\lambda \in (0,\infty)$. There is therefore $n_0 \in \mathbb{N}$ such that for every $n > n_0$ we have

$$\frac{d\left(x, J_{\lambda_0}^n x\right)}{\sqrt{\lambda_0}} \le \frac{d\left(x, J_{\lambda_0} x\right)}{\sqrt{\lambda_0}} + \varepsilon,$$

and hence

$$d\left(x, J_{\lambda_0}^n x\right) \le \sqrt{\lambda_0} \left|\partial f\right|(x) + \varepsilon\sqrt{\lambda_0} < 2\varepsilon.$$

Next choose $k_0 \in \mathbb{N}$ such that

$$\frac{t}{\sqrt{2k_0}} \left|\partial f\right|(x) < \varepsilon,$$

and simultaneously,

$$\frac{t}{\sqrt{2k_0}} \frac{d\left(x, J_{\lambda_0}^n x\right)}{\lambda_0} < \varepsilon,$$

for every $n > n_0$. Then we can find $n_1 > n_0$ so that for every $n > n_1$ we have

$$d\left(\left(J_{\frac{t}{k_0}}^n\right)^{(k_0)} x, \left(J_{\frac{t}{k_0}}\right)^{(k_0)} x\right) < \varepsilon.$$

Altogether we have

$$d\left(S_t^n x, S_t x\right) < 7\varepsilon,$$

for each $n > n_1$.

Step 3: Let finally $x \in \overline{\operatorname{dom} f}$. Since $\operatorname{dom} |\partial f|$ is dense in $\overline{\operatorname{dom} f}$ by Lemma 5.1.3, there exists, for each $\delta > 0$, a point $y \in \operatorname{dom} |\partial f|$ such that $d(x, y) < \delta$. Then

$$d\left(S_t^n x, S_t x\right) \leq d\left(S_t^n x, S_t^n y\right) + d\left(S_t^n y, S_t y\right) + d\left(S_t y, S_t x\right) < 2\delta + d\left(S_t^n y, S_t y\right),$$

which finishes the proof. □

Question 5.2.10. Is the convergence in (5.2.29) uniform with respect to t on each bounded time interval?

5.3 Lie–Trotter–Kato formula

Let $(\mathcal{H}, d)$ be a Hadamard space. In this section we consider a function $f : \mathcal{H} \to (-\infty, \infty]$ of the form

$$f := \sum_{n=1}^{N} f_n, \tag{5.3.30}$$

where $f_n : \mathcal{H} \to (-\infty, \infty]$ are convex lsc functions, $n = 1, \ldots, N$ and $N \in \mathbb{N}$. The Lie–Trotter–Kato formula tells us that the semigroup of f can be approximated by the resolvents of the components f_n. Functions of the type (5.3.30) appeared for instance in Examples 2.2.5 and 2.2.6 and will be revisited in (6.3.15) and (8.3.2).

Before stating the main result in Theorem 5.3.7, we need several auxiliary results. Define first a resolvent of a *family* of nonexpansive mappings, extending Definition 4.2.1.

Definition 5.3.1. Let $(\mathcal{H}, d)$ be a Hadamard space and $F_\rho : \mathcal{H} \to \mathcal{H}$ be a nonexpansive map for each $\rho \in (0, \infty)$. Given $\lambda \in (0, \infty)$ and $x \in \mathcal{H}$, the map

$$y \mapsto \frac{1}{1 + \frac{\lambda}{\rho}} x + \frac{\frac{\lambda}{\rho}}{1 + \frac{\lambda}{\rho}} F_\rho y, \quad y \in \mathcal{H}, \tag{5.3.31}$$

is a contraction with Lipschitz constant $\frac{\lambda/\rho}{1+(\lambda/\rho)}$ and hence has a unique fixed point, which will be denoted $R_{\lambda,\rho}(x)$. The mapping $x \mapsto R_{\lambda,\rho}(x)$ is called the *resolvent* of the family (F_ρ).

Definition 5.3.2. Let $(\mathcal{H}, d)$ be a Hadamard space and $F_\rho : \mathcal{H} \to \mathcal{H}$ be a nonexpansive map for each $\rho \in (0, \infty)$. The *semigroup* of the family (F_ρ) is the mapping

$$T_t^\rho x := \lim_{j \to \infty} R_{\frac{t}{j},\rho}^{(j)} x, \quad x \in \mathcal{H}.$$

where $t \in [0, \infty)$.

If we replace F by F_ρ in Theorem 4.3.3 and its proof, we can see that (T_t^ρ) is a well-defined strongly continuous semigroup of nonexpansive mappings and satisfies

$$d\left(T_t^\rho x, T_s^\rho x\right) \le 2\frac{d\left(x, F_\rho x\right)}{\rho}|t-s|, \quad x \in \mathcal{H},$$

for every $s, t \in [0, \infty)$, and

$$d\left(T_t^\rho x, R_{\frac{t}{j},\rho}^{(j)} x\right) \le \frac{2t}{\sqrt{j}}\frac{d\left(x, F_\rho x\right)}{\rho}, \quad x \in \mathcal{H}, \tag{5.3.32}$$

for every $t \in [0, \infty)$ and $j \in \mathbb{N}$.

Lemma 5.3.3. *Let $F_\rho \colon \mathcal{H} \to \mathcal{H}$ be a nonexpansive map for each $\rho \in (0, \infty)$. The semigroup of the family (F_ρ) satisfies*

$$d\left(T_t^\rho x, F_\rho^{(k)} x\right) \le 2d(x, Fx)\sqrt{\left(k - \frac{t}{\rho}\right)^2 + \frac{t}{\rho}}, \quad x \in \mathcal{H},$$

for every $t \in [0, \infty)$ and $k \in \mathbb{N}_0$.

Proof. Lemma 4.3.5 gives the estimate

$$d\left(T_t^1 x, F_\rho^{(k)} x\right) \le 2d(x, Fx)\sqrt{(k-t)^2 + t}, \quad x \in \mathcal{H}.$$

Next observe that $T_t^\rho = T_{\frac{t}{\rho}}^1$, because $R_{\lambda,\rho} = R_{\frac{\lambda}{\rho},1}$. Combining these two facts finishes the proof. □

It is useful to set up some notation.

Notation 5.3.4. Let $f \colon \mathcal{H} \to (-\infty, \infty]$ be a function of the form (5.3.30). Denote its resolvent J_λ and semigroup (S_t), respectively. The resolvent of the function f_n is denoted J_λ^n, for each $n = 1, \ldots, N$. At the same time, there will be a family of nonexpansive maps $F_\rho \colon \mathcal{H} \to \mathcal{H}$ and we will use its resolvents $R_{\lambda,\rho}$ and semigroup (T_t^ρ) defined in Definitions 5.3.1 and 5.3.2, respectively.

The proof of the Lie–Trotter–Kato formula relies upon the next two approximation results (Theorems 5.3.5 and 5.3.6) which may also be of independent interest. It would be more economical to state these two theorems as a single result, but we choose not to do so for the sake of clarity.

Theorem 5.3.5. *Let $(\mathcal{H}, d)$ be a Hadamard space and $f \colon \mathcal{H} \to (-\infty, \infty]$ be of the form (5.3.30). We use Notation 5.3.4 and put $F_\rho := J_\rho^N \circ \cdots \circ J_\rho^1$. If we have*

$$J_\lambda x = \lim_{\rho \to 0+} R_{\lambda,\rho} x,$$

for every $x \in \overline{\operatorname{dom} f}$ *and* $\lambda \in [0,\infty)$, *then,*

$$S_t x = \lim_{\rho \to 0+} T_t^\rho x\,, \tag{5.3.33}$$

for every $x \in \overline{\operatorname{dom} f}$ *and* $t \in [0,\infty)$.

Proof. Step 1: Let us first assume that $x \in \operatorname{dom} |\partial f|$. By the triangle inequality,

$$\begin{aligned} d\left(T_t^\rho x, S_t x\right) &\le d\left(T_t^\rho x, T_t^\rho R_{\lambda,\rho} x\right) + d\left(T_t^\rho R_{\lambda,\rho} x, S_t x\right) \\ &\le d\left(x, R_{\lambda,\rho} x\right) + d\left(T_t^\rho R_{\lambda,\rho} x, S_t x\right)\,, \end{aligned} \tag{5.3.34}$$

and the second term on the right hand side can be further estimated

$$\begin{aligned} d\left(T_t^\rho R_{\lambda,\rho} x, S_t x\right) \le\; & d\left(T_t^\rho R_{\lambda,\rho} x, R^{(j)}_{\frac{t}{j},\rho} R_{\lambda,\rho} x\right) + d\left(R^{(j)}_{\frac{t}{j},\rho} R_{\lambda,\rho} x, R^{(j)}_{\frac{t}{j},\rho} x\right) \\ &+ d\left(R^{(j)}_{\frac{t}{j},\rho} x, J^{(j)}_{\frac{t}{j}} x\right) + d\left(J^{(j)}_{\frac{t}{j}} x, S_t x\right)\,. \end{aligned}$$

The error estimate in (5.3.32) gives

$$d\left(T_t^\rho R_{\lambda,\rho} x, R^{(j)}_{\frac{t}{j},\rho} R_{\lambda,\rho} x\right) \le \frac{2t}{\sqrt{j}} \frac{d\left(R_{\lambda,\rho} x, F_\rho R_{\lambda,\rho} x\right)}{\rho} = \frac{2t}{\lambda\sqrt{j}} d\left(x, R_{\lambda,\rho} x\right)\,,$$

and applying also (5.1.5), inequality (5.3.34) becomes

$$\begin{aligned} d\left(T_t^\rho x, S_t x\right) \le\; & 2d\left(x, R_{\lambda,\rho} x\right) + \frac{2t}{\lambda\sqrt{j}} d\left(x, R_{\lambda,\rho} x\right) \\ &+ d\left(R^{(j)}_{\frac{t}{j},\rho} x, J^{(j)}_{\frac{t}{j}} x\right) + \frac{t}{\sqrt{2j}} |\partial f|(x)\,, \end{aligned} \tag{5.3.35}$$

for each $j \in \mathbb{N}$. At the next step, we will show that all the terms on the right hand side tend to 0 as $\rho \to 0$.

Step 2: Choose $\varepsilon > 0$ and find $\lambda_0 \in (0,1)$ such that $\lambda_0 |\partial f|(x) < \varepsilon$. By the assumptions and (5.1.3), we have

$$\lim_{\rho \to 0+} d\left(x, R_{\lambda,\rho} x\right) = d\left(x, J_\lambda x\right) \le \lambda |\partial f|(x)\,.$$

for each $\lambda > 0$, and hence there exists $\rho_0 > 0$ such that

$$d\left(x, R_{\lambda_0,\rho} x\right) \le d\left(x, J_{\lambda_0} x\right) + \varepsilon \le \lambda_0 |\partial f|(x) + \varepsilon < 2\varepsilon\,,$$

for every $\rho \in (0, \rho_0)$. Next find $j_0 \in \mathbb{N}$ such that

$$\frac{t}{\sqrt{2 j_0}} |\partial f|(x) < \varepsilon\,,$$

and simultaneously

$$\frac{2t}{\lambda_0\sqrt{j_0}}d\left(x, R_{\lambda_0,\rho}x\right) < \varepsilon\,,$$

for every $\rho \in (0, \rho_0)$. Finally, by the assumptions we can find $\rho_1 \in (0, \rho_0)$ such that

$$d\left(R^{(j_0)}_{\frac{t}{j_0},\rho}x, J^{(j_0)}_{\frac{t}{j_0}}x\right) < \varepsilon\,.$$

for every $\rho \in (0, \rho_1)$. This finishes the proof of (5.3.33) for $x \in \operatorname{dom}|\partial f|$.

Step 3: Let now $x \in \overline{\operatorname{dom} f}$ and $\delta > 0$. By Lemma 5.1.3 there exists $y \in \operatorname{dom}|\partial f|$ with $d(x, y) < \delta$. Then

$$\begin{aligned} d\left(T_t^\rho x, S_t x\right) &\le d\left(T_t^\rho x, T_t^\rho y\right) + d\left(T_t^\rho y, S_t y\right) + d\left(S_t y, S_t x\right) \\ &\le 2d(x, y) + d\left(T_t^\rho y, S_t y\right)\,, \end{aligned}$$

which gives (5.3.33) for $x \in \overline{\operatorname{dom} f}$. □

Theorem 5.3.6. *Let $(\mathcal{H}, d)$ be a Hadamard space and $f\colon \mathcal{H} \to (-\infty, \infty]$ be of the form (5.3.30). We use Notation 5.3.4. Let $R_{\lambda,\rho}$ be the resolvent of the map $F_\rho := J_\rho^N \circ \cdots \circ J_\rho^1$. If we have*

$$J_\lambda x = \lim_{\rho\to 0+} R_{\lambda,\rho}x\,,$$

for every $x \in \overline{\operatorname{dom} f}$ and $\lambda \in [0, \infty)$, then,

$$S_t x = \lim_{j\to\infty} \left(J^N_{\frac{t}{j}} \circ \cdots \circ J^1_{\frac{t}{j}}\right)^{(j)} x\,,$$

for every $x \in \overline{\operatorname{dom} f}$ and $t \in [0, \infty)$.

Proof. Assume that $x \in \operatorname{dom}|\partial f|$. We have

$$d\left(S_t x, F^{(j)}_{\frac{t}{j}}x\right) \le d\left(S_t x, T_t^{\frac{t}{j}}x\right) + d\left(T_t^{\frac{t}{j}}x, F^{(j)}_{\frac{t}{j}}x\right)\,,$$

for $j \in \mathbb{N}$, and the first term on the right hand side converges to 0 as $j \to \infty$ by Theorem 5.3.5. The second term can be estimated

$$\begin{aligned} d\left(T_t^{\frac{t}{j}}x, F^{(j)}_{\frac{t}{j}}x\right) &\le d\left(T_t^{\frac{t}{j}}x, T_t^{\frac{t}{j}}R_{\lambda,\frac{t}{j}}x\right) + d\left(T_t^{\frac{t}{j}}R_{\lambda,\frac{t}{j}}x, F^{(j)}_{\frac{t}{j}}R_{\lambda,\frac{t}{j}}x\right) + d\left(F^{(j)}_{\frac{t}{j}}R_{\lambda,\frac{t}{j}}x, F^{(j)}_{\frac{t}{j}}x\right) \\ &\le 2d\left(x, R_{\lambda,\frac{t}{j}}x\right) + d\left(T_t^{\frac{t}{j}}R_{\lambda,\frac{t}{j}}x, F^{(j)}_{\frac{t}{j}}R_{\lambda,\frac{t}{j}}x\right)\,, \end{aligned}$$

for $j \in \mathbb{N}$ and $\lambda > 0$. By Lemma 5.3.3 we obtain

$$d\left(T_t^{\frac{t}{j}}R_{\lambda,\frac{t}{j}}x, F^{(j)}_{\frac{t}{j}}R_{\lambda,\frac{t}{j}}x\right) \le 2\sqrt{j}d\left(R_{\lambda,\frac{t}{j}}x, F_{\frac{t}{j}}R_{\lambda,\frac{t}{j}}x\right) = \frac{2t}{\lambda\sqrt{j}}d\left(x, R_{\lambda,\frac{t}{j}}x\right)\,,$$

which can be made arbitrarily small by taking $j \in \mathbb{N}$ big enough. Choose now $\varepsilon > 0$ and find $\lambda_0 \in (0,1)$ such that $\lambda_0 |\partial f|(x) < \varepsilon$. By the assumptions and (5.1.3), we have

$$\lim_{\rho\to 0+} d\left(x, R_{\lambda_0,\rho}x\right) = d\left(x, J_{\lambda_0}x\right) \leq \lambda_0 |\partial f|(x)\,.$$

and hence there exists $j_0 \in \mathbb{N}$ such that

$$2d\left(x, R_{\lambda_0,\frac{t}{j_0}}x\right) \leq 2d\left(x, J_{\lambda_0}x\right) + \varepsilon \leq 2\lambda_0|\partial f|(x) + \varepsilon < 3\varepsilon\,,$$

which proves the statement for $x \in \operatorname{dom}|\partial f|$.

Using an easy approximation argument like in the proof of Theorem 5.3.5, the statement holds also for $x \in \overline{\operatorname{dom} f}$. □

We are now ready to present the Lie–Trotter–Kato product formula, which says that the semigroup of the function f of the form (5.3.30) can be approximated by the resolvents of the components f_n, with $n = 1, \dots, N$.

Theorem 5.3.7 (Lie–Trotter–Kato formula). *Let $(\mathcal{H}, d)$ be a Hadamard space and $f\colon \mathcal{H} \to (-\infty, \infty]$ be of the form (5.3.30). We use Notation 5.3.4. Then we have*

$$S_t x = \lim_{j\to\infty} \left(J^N_{\frac{t}{j}} \circ \cdots \circ J^1_{\frac{t}{j}}\right)^{(j)} x\,, \tag{5.3.37}$$

for every $t \in [0, \infty)$ and $x \in \overline{\operatorname{dom} f}$.

Proof. Step 1: By Theorem 5.3.6, it suffices to show $R_{\lambda,\rho}x \to J_\lambda x$ as $\rho \to 0$, where $R_{\lambda,\rho}$ is the resolvent of $F_\rho := J^N_\rho \circ \cdots \circ J^1_\rho$. Put $x_0(\rho) := R_{\lambda,\rho}x$ and $x_n(\rho) := J^n_\rho x_{n-1}(\rho)$ for $n = 1, \dots, N$. By the definition of $R_{\lambda,\rho}$ we have

$$x_0(\rho) = \frac{1}{1+\frac{\lambda}{\rho}}x + \frac{\frac{\lambda}{\rho}}{1+\frac{\lambda}{\rho}}x_N(\rho)\,, \tag{5.3.38}$$

and consequently,

$$d\left(x, x_N(\rho)\right) = \frac{\rho+\lambda}{\lambda} d\left(x, x_0(\rho)\right)\,, \tag{5.3.39}$$

together with,

$$d\left(x_0(\rho), x_N(\rho)\right) = \frac{\rho}{\lambda} d\left(x, x_0(\rho)\right)\,. \tag{5.3.40}$$

Step 2: Applying Lemma 2.2.23 for each f_n, with $n = 1, \dots, N$, and summing the resulting inequalities up, we arrive at

$$\begin{aligned} 2\rho f(v) \geq 2\rho \sum_{n=1}^{N} f_n\left(x_n(\rho)\right) + d\left(x_N(\rho), v\right)^2 - d\left(x_0(\rho), v\right)^2 \\ + \sum_{n=1}^{N} d\left(x_{n-1}(\rho), x_n(\rho)\right)^2\,, \end{aligned} \tag{5.3.41}$$

for every $v \in \operatorname{dom} f$. Inequality (1.2.2) yields

$$\frac{\frac{\lambda}{\rho}}{1+\frac{\lambda}{\rho}} d\left(v, x_N(\rho)\right)^2 \geq d\left(v, x_0(\rho)\right)^2 - \frac{1}{1+\frac{\lambda}{\rho}} d(v,x)^2 + \frac{\frac{\lambda}{\rho}}{\left(1+\frac{\lambda}{\rho}\right)^2} d\left(x, x_N(\rho)\right)^2 . \tag{5.3.42}$$

Combining this inequality with (5.3.39) and (5.3.41) gives after some elementary calculations that

$$2\lambda f(v) \geq 2\lambda \sum_{n=1}^{N} f_n\left(x_n(\rho)\right) + d\left(x_0(\rho), x\right)^2 + d\left(x_0(\rho), v\right)^2 - d(x,v)^2 . \tag{5.3.43}$$

for every $v \in \operatorname{dom} f$.

Step 3: Fix now a sequence $\rho_j \to 0$. We will show that, for every $n = 0, 1, \ldots, N$, the sequence $(x_n(\rho_j))_j$ is bounded. To this end, apply Lemma 2.2.13 to obtain $\alpha, \beta \in \mathbb{R}$ such that

$$f_n\left(x_n\left(\rho_j\right)\right) \geq \alpha + \beta d\left(x, x_n\left(\rho_j\right)\right) ,$$

for every $n = 1, \ldots, N$ and $j \in \mathbb{N}$. Since, by (1.1), we have

$$d\left(x_0\left(\rho_j\right), v\right)^2 \geq \frac{1}{2} d\left(x_0\left(\rho_j\right), x\right)^2 - d(x,v)^2 ,$$

inequality (5.3.43) yields

$$\lambda f(v) \geq \lambda\left[N\alpha + \beta \sum_{n=1}^{N} d\left(x, x_n\left(\rho_j\right)\right)\right] + \frac{3}{4} d\left(x_0\left(\rho_j\right), x\right)^2 - d(x,v)^2 . \tag{5.3.44}$$

for every $v \in \operatorname{dom} f$. The case $\beta \geq 0$ is easy, we therefore assume $\beta < 0$. Next, observe that

$$\begin{aligned} d\left(x, x_n\left(\rho_j\right)\right) &\leq d\left(x, J_{\rho_j}^n x\right) + d\left(J_{\rho_j}^n x, J_{\rho_j}^n J_{\rho_j}^{n-1} x\right) + \cdots + d\left(J_{\rho_j}^n \cdots J_{\rho_j}^1 x, x_n\left(\rho_j\right)\right) \\ &\leq \sum_{k=1}^{n} d\left(x, J_{\rho_j}^k x\right) + d\left(x, x_0\left(\rho_j\right)\right) , \\ &\leq L + d\left(x, x_0\left(\rho_j\right)\right) \end{aligned} \tag{5.3.45}$$

for some $L > 0$ and every $n = 1, \ldots, N$ and $j \in \mathbb{N}$. Plugging this inequality into (5.3.44) gives that $(x_0(\rho_j))_n$ is bounded and by (5.3.45) we get that the sequence $(x_n(\rho_j))_j$ is bounded also for $n = 1, \ldots, N$.

Then Lemma 2.2.12 yields that the sequence $(f_n(x_n(\rho_j)))_j$ is bounded from below and inequality (5.3.43) implies that it is also bounded from above.

Step 4: Consider (5.3.41) with $\rho := \rho_j$ and take the limit $j \to \infty$ to obtain via (5.3.40) that

$$\lim_{j\to\infty} \sum_{n=1}^{N} d\left(x_{n-1}\left(\rho_j\right), x_n\left(\rho_j\right)\right)^2 = 0 . \tag{5.3.46}$$

Let $z \in \mathcal{H}$ be a weak cluster point of $x_0(\rho_j)$ and $x_0(\rho_{j_p})$ be a sequence weakly converging to z. Recall that the existence of a weak cluster point was guaranteed by Proposition 3.1.2. By (5.3.46), also $x_j(\rho_{j_p})$ weakly converges to z, for every $n = 1, \dots, N$. Consequently, $z \in \overline{\operatorname{dom} f}$ due to Lemma 3.2.1.

Consider next inequality (5.3.43) with $\rho := \rho_{j_p}$ and take the limit $p \to \infty$. Since the functions f_n, with $n = 1, \dots, N$, are weakly lsc by Lemma 3.2.3, we obtain $z = J_\lambda x$. In particular, $z \in \operatorname{dom} f$. Since z was an arbitrary weak cluster point of $x_0(\rho_j)$, we get $x_0(\rho_j) \xrightarrow{w} J_\lambda x$. Applying once again (5.3.43) with $v := J_\lambda x$ gives $x_0(\rho_j) \to J_\lambda x$ as $j \to \infty$. This finishes the proof of (5.3.37). □

Question 5.3.8. Is the convergence in (5.3.37) uniform with respect to t on each bounded time interval?

Exercises

Exercise 5.1. Find a convex lsc function $f\colon \mathbb{R} \to (-\infty, \infty]$ such that $\operatorname{dom} |\partial f| \neq \operatorname{dom} f$.

Exercise 5.2. Let $v\colon (0,1) \to \mathbb{R}$ be a locally absolutely continuous function and $D\colon (0,1)^2 \to \mathbb{R}$ be a map satisfying

$$\left|D(s,t) - D\left(s',t\right)\right| \le \left|v(s) - v\left(s'\right)\right|, \quad \left|D(s,t) - D\left(s,t'\right)\right| \le \left|v(t) - v\left(t'\right)\right|,$$

for every $s, s', t, t' \in (0,1)$. Show that the function $t \mapsto D(t,t)$ is locally absolutely continuous on $(0,1)$ and

$$\frac{\mathrm{d}}{\mathrm{d}t} D(t,t) \le \limsup_{h \to 0+} \frac{D(t,t) - D(t-h,t)}{h} + \limsup_{h \to 0+} \frac{D(t,t+h) - D(t,t)}{h},$$

for almost every $t \in (0,1)$.
Hint. See [8, Lemma 4.3.4]. □

Exercise 5.3. Let $f\colon \mathcal{H} \to (-\infty, \infty]$ be convex lsc and $x \in \mathcal{H}$. Show that for every $t \in [0, \infty)$ we have

$$\sup_{n \in \mathbb{N}} d\left(x, J_{\frac{t}{n}}^{(n)} x\right) < \infty$$

Hint. First use (5.1.3) to get the claim for $x \in \operatorname{dom} |\partial f|$ and then apply the fact that $\operatorname{dom} |\partial f|$ is dense in $\overline{\operatorname{dom} f}$ along with the nonexpansiveness of the resolvent, which gives the claim for $x \in \overline{\operatorname{dom} f}$. Finally, use the limit behavior of the resolvent at 0 from Proposition 2.2.26. □

Exercise 5.4 (Gradient flows of strongly convex functions). Prove Proposition 5.1.15.

Exercise 5.5 (Strong convergence). Give an elementary proof of the fact that if $f\colon \mathcal{H} \to (-\infty, \infty]$ in Theorem 5.1.16 is *strongly* convex, then the gradient flow converges *strongly* to a minimizer.

Exercise 5.6. Let $(\mathcal{H}, d)$ be a Hadamard space and $f^n\colon \mathcal{H} \to (-\infty, \infty]$ be a sequence of functions which Γ-converges to a function $f\colon \mathcal{H} \to (-\infty, \infty]$. Show that f is convex whenever f^n are such.

Exercise 5.7. Let (X, d) be a complete metric space and $f^n\colon X \to (-\infty, \infty]$ be a sequence of functions which Γ-converges to a function $f\colon X \to (-\infty, \infty]$. Show that f is lsc.

Bibliographical remarks

A classical reference for problem (5.0.2) is H. Brezis' book [46]. The study of gradient flows in Hadamard spaces was initiated independently by U. Mayer [139] and J. Jost [105], who both obtained the main existence result (5.1.4) and the semigroup property (Proposition 5.1.8) based on the celebrated Crandall–Liggett construction in linear spaces [68]. There had been however earlier results of this type in a special case [173, 175]. On the other hand a significant part of this theory works in more general metric spaces and is now a hot topic in analysis [8]. Our exposition in Section 5.1 follows the development in [8], which we adapt into Hadamard spaces. A considerable number of these results were already established in [139], namely, the Hölder and Lipschitz property of the flow (Proposition 5.1.10) coming from [139, Theorems 2.2 and 2.9], and including the function (5.1.13). Furthermore, the properties of a gradient flow in Theorem 5.1.13 are from [139] and in the case of (5.1.17), we follow the original proof [139, Theorem 2.14]. The proof of (5.1.16) comes from [9, p. 84]. Large time behavior of a gradient flow (Theorem 5.1.16) comes from [19]. Strongly convex functions were known to produce strongly convergent gradient flow already in [139, Lemma 1.7]. For the Hilbert space case, see [31, Theorem 27.1 (iii)]. The existence of a gradient flow in Hadamard spaces and its basic properties can be alternatively established in a more elementary way [7, Theorem 4.25].

Theorems 5.2.4 and 5.2.9 come from [17], but earlier results on Mosco and Γ-convergences in Hadamard space were obtained already in [105, 124]. For linear counterparts of these two theorems, see [12, Theorem 3.26], [13, Théorème 1.2] and [46, Theorems 3.16 and 4.2]. Closely related results can be found for instance in [49, 68]. The claim in the proof of Theorem 5.2.4 was in the context of Hilbert spaces proved in [13, Lemme 1.5]. For the details on the Frolík–Wijsman convergence, see [87, 199, 203]. A Banach space version of Corollary 5.2.8 (ii) appears in the paper [101] by M. Israel, Jr. and S. Reich. Question 5.2.10 was motivated by [46, Theorems 3.16 and 4.2].

Theorem 5.3.7 is due to I. Stojkovic [189, Theorem 4.4]. Our proof comes from [18]. The Lie–Trotter–Kato formula in Banach spaces is a classical part of functional analysis. It originated in seminal works by H. Brezis and A. Pazy [48, 49], P. Chernoff [60], T. Kato [107], I. Miyadera and S. Ôharu [144], S. Reich [168–170], H. Trotter [197, 198] and others. Theorem 5.3.7 is particularly close to [108]. Question 5.3.8 is natural with regard to the above mentioned Hilbert space results. For the details, see also the monographs [46, 61, 79, 164]. We also note that Theorem 5.3.5 was proved in [189, Theorem 3.12] with the additional claim that the semigroup convergence is *uniform* in t on compact intervals. I was however unable to understand the argument in the proof leading to such a claim and Theorem 5.3.5 is therefore stated here without *uniform* convergence. Consequently, we do not have uniform convergence in Theorem 5.3.6, which is taken from [189, Theorem 3.13], and ultimately we pose Question 5.3.8. A very nice application of the Lie–Trotter–Kato formula in Hadamard spaces was discovered by L. Kovalev in [122].

We should like to also mention a semigroup associated to a nonlinear Markov operator in a Hadamard space, which was obtained by a formula similar to (5.1.4) in [192, Section 8].

I would like to thank Jeffrey Streets for a discussion about Calabi flows and the Donaldson conjecture.

6 Convex optimization algorithms

After being concerned with convex analysis in the previous chapters, we now turn to a few topics in convex optimization. That is, computational and algorithmic questions about convex sets and convex functions will be studied. Since the optimization problems under consideration are typically impossible to solve exactly, we develop *approximation* methods, which means that a sequence of points is generated by an algorithm and its convergence to a desired exact solution is proved. In particular, the algorithms studied in this chapter work in infinitely many steps.

We present several optimization methods which are classical in Hilbert spaces and show that they function in Hadamard spaces equally well. In Section 8.3 we will see that such generalizations are not only of theoretical interest, but can be important in real world applications.

It is well worth pointing out that all the algorithms rely on the fact that a metric projection on a closed convex set is a nonexpansive mapping (Theorem 2.1.12). Also Fejér monotonicity (Definition 3.2.5) plays a prominent role in the proofs of convergence.

6.1 Convex feasibility problems

In many situations in optimization, we are to minimize a function on a set C which is given as an intersection of (finitely many) closed convex sets and before running a minimization algorithm itself, one needs to find a feasible solution, that is, a point from C. This task is in the literature called a *convex feasibility problem*. Here we will consider the following case: given two closed convex subsets A and B of a Hadamard space $(\mathcal{H}, d)$ such that $A \cap B \neq \emptyset$, we are to find a point $x \in A \cap B$. We shall present two approximation algorithms, namely the alternating projection method and the averaged projection method, each of which generates a sequence of points $(x_n) \subset \mathcal{H}$ converging to a point $x \in A \cap B$.

It turns out that in order to get *strong* convergence of such a sequence (x_n), we need additional assumptions. The following regularity conditions mean that if a point is close to both sets A and B, then it is close to their intersection $A \cap B$. Under even stronger assumptions on the sets, one obtains *linear* convergence of the algorithm; see (3.2.4) to recall this type of convergence.

Regularity of two intersecting sets

We say that $A, B \subset \mathcal{H}$ are *boundedly regular* if for every bounded set $S \subset \mathcal{H}$ and $\varepsilon > 0$ there exists $\delta > 0$ such that $d(x, A \cap B) < \varepsilon$ whenever $x \in S$ and $\max\{d(x, A), d(x, B)\} \leq \delta$.

The sets $A, B \subset \mathcal{H}$ are *boundedly linearly regular* if for every bounded set $S \subset \mathcal{H}$ there exists $\kappa > 0$ such that we have

$$d(x, A \cap B) \leq \kappa \max\{d(x, A), d(x, B)\},$$

for each $x \in S$.

Finally, we say that $A, B \subset \mathcal{H}$ are *linearly regular* if there exists $\kappa > 0$ such that we have

$$d(x, A \cap B) \leq \kappa \max\{d(x, A), d(x, B)\},$$

for each $x \in \mathcal{H}$.

Alternating projection method

Now we are ready to introduce the alternating projection method, a well-known optimization algorithm in Hilbert spaces, which can be traced back to von Neumann. Let $(\mathcal{H}, d)$ be a Hadamard space and $A, B \subset \mathcal{H}$ convex closed subsets such that $A \cap B \neq \emptyset$. Choose a starting point $x_0 \in \mathcal{H}$ and define

$$x_{2n-1} := P_A x_{2n-2}, \quad x_{2n} := P_B x_{2n-1}, \quad n \in \mathbb{N}. \tag{6.1.1}$$

This sequence is sometimes referred to as the *alternating sequence*.

Theorem 6.1.1 (Alternating projections). *Let $(x_n) \subset \mathcal{H}$ be the sequence generated by* (6.1.1). *Then the following holds:*
(i) *The sequence (x_n) weakly converges to a point $x \in A \cap B$.*
(ii) *If A and B are boundedly regular, then $x_n \to x$.*
(iii) *If A and B are boundedly linearly regular, then $x_n \to x$ linearly.*
(iv) *If A and B are linearly regular, then $x_n \to x$ linearly with a rate independent of the starting point.*

Proof. By (2.1.1) the sequence (x_n) is Fejér monotone with respect to $A \cap B$. We claim that

$$\max\left\{d(x_n, A)^2, d(x_n, B)^2\right\} \leq d(x_n, A \cap B)^2 - d(x_{n+1}, A \cap B)^2, \tag{6.1.2}$$

for every $n \in \mathbb{N}$. Indeed, fix $n \in \mathbb{N}$ and without loss of generality assume $x_n \in A$ and $x_{n+1} \notin A \cap B$. Inequality (2.1.1) reads

$$d(x_n, P_{A\cap B}(x_n))^2 \geq d(x_n, x_{n+1})^2 + d(P_{A\cap B}(x_n), x_{n+1})^2,$$
$$d(x_n, A \cap B)^2 \geq d(x_n, B)^2 + d(A \cap B, x_{n+1})^2,$$

which yields (6.1.2). Now, by Fejér monotonicity, Proposition 3.2.6 (ii) and (6.1.2) we get

$$\max\{d(x_n, A), d(x_n, B)\} \to 0, \quad \text{as } n \to \infty. \tag{6.1.3}$$

Let us prove (i). Since (x_n) is bounded, it has a weak cluster point $x \in \mathcal{H}$. Take a subsequence (x_{n_k}) which weakly converges to x. Using Corollary 3.2.4 and (6.1.3), we have $d(x, A) = d(x, B) = 0$. Hence $x \in A \cap B$ and we conclude, by Proposition 3.2.6 (iii), that $x_n \xrightarrow{w} x$.

As for (ii), the bounded regularity of A and B along with (6.1.3) gives $d(x_n, A \cap B) \to 0$ as $n \to \infty$. Applying Proposition 3.2.6 (iv) yields (ii).

To prove (iii), recall that (x_n) is bounded. Hence, by bounded linear regularity, there exists $\kappa > 0$ such that

$$d(x_n, A \cap B) \leq \kappa \max\{d(x_n, A), d(x_n, B)\} ,$$

for every $n \in \mathbb{N}$. Using (6.1.2), we arrive at

$$d^2(x_n, A \cap B) \leq \kappa^2 \left[d^2(x_n, A \cap B) - d^2(x_{n+1}, A \cap B)\right] ,$$

$$d(x_{n+1}, A \cap B) \leq \sqrt{1 - \frac{1}{\kappa^2}}\, d(x_n, A \cap B).$$

Applying Proposition 3.2.6 (v) finishes the proof of (iii).

Finally, the proof of (iv) is similar to that one of (iii). □

It is natural to ask whether it is possible to obtain the strong convergence of (6.1.1) in general. This was a longstanding problem in Hilbert spaces, which eventually turned out to have a negative answer.

Example 6.1.2. There exist a hyperplane $A \subset \ell_2$, a closed convex cone $B \subset \ell_2$ and a point $x_0 \in \ell_2$ such that the sequence generated by (6.1.1) from the starting point x_0 converges weakly to a point in $A \cap B$, but not in norm [100, 138].

On the other hand already John von Neumann showed that if H is a Hilbert space and $A, B \subset H$ are its closed subspaces, then for any starting point $x_0 \in H$, the sequence defined by (6.1.1) converges *in norm* to a point from $A \cap B$.

There exist surprisingly nice sets $A, B \subset \ell_2$ for which we do not know whether the convergence of the alternating sequence (6.1.1) is strong.

Question 6.1.3. Let $A \subset \ell_2$ be a closed (affine) subspace and

$$B := \{(x_n) \in \ell_2 \colon x_n \geq 0 \text{ for every } n \in \mathbb{N}\}$$

be the positive cone. Assume $A \cap B \neq \emptyset$ and $x_0 \in \ell_2$. Does the sequence generated by (6.1.1) converge *strongly* to a point from $A \cap B$?

Averaged projection method

Another popular method for solving convex feasibility problems is the *averaged projection method*. Let again A and B be closed convex subsets of a Hadamard space

$(\mathcal{H}, d)$ with $A \cap B \neq \emptyset$. Choose a starting point $x_0 \in \mathcal{H}$ and put

$$x_n := \frac{P_A x_{n-1} + P_B x_{n-1}}{2}, \quad n \in \mathbb{N}. \tag{6.1.5}$$

That is, at each step, the averaged projection method updates x_{n-1} to x_n by projecting x_{n-1} on A and B, respectively, and taking the midpoint of the two projections $P_A x_{n-1}$ and $P_B x_{n-1}$.

Theorem 6.1.4 (Averaged projections). *Let (x_n) be a sequence generated by (6.1.5). Then the following holds:*
(i) *The sequence (x_n) weakly converges to a point $x \in A \cap B$.*
(ii) *If A and B are boundedly regular, then $x_n \to x$.*
(iii) *If A and B are boundedly linearly regular, then $x_n \to x$ linearly.*
(iv) *If A and B are linearly regular, then $x_n \to x$ linearly with a rate independent of the starting point.*

Proof. The proof is rather similar to that of Theorem 6.1.1 so we will be brief. The sequence (x_n) is Fejér with respect to $A \cap B$. Let $c \in \mathcal{H}$ be a weak cluster point of (x_n). Since

$$\begin{aligned} d(x_n, A)^2 &\leq d(x_n, A \cap B)^2 - d\left(P_A x_n, A \cap B\right)^2 \\ d(x_n, A)^2 &\leq d(x_n, A \cap B)^2 - d\left(x_{n+1}, A \cap B\right)^2, \end{aligned}$$

and after interchanging A and B also

$$d(x_n, B)^2 \leq d(x_n, A \cap B)^2 - d\left(x_{n+1}, A \cap B\right)^2,$$

which together with the weak lower semicontinuity of the metric yields $c \in A \cap B$. By Proposition 3.2.6 (iii), we finally get $x_n \xrightarrow{w} x$. This proves (i). The remainder is similar to the proof of Theorem 6.1.1 and we leave it as Exercise 6.1. □

In general, the convergence in Theorem 6.1.4 is not strong [32].

Remark 6.1.5. Convex feasibility problems can be also formulated in the language of fixed point theory. Note that $x \in A \cap B$ if and only if $(\frac{P_A + P_B}{2})x = x$. And likewise for $P_B \circ P_A$, more precisely, $x \in A \cap B$ if and only if $(P_B \circ P_A)x = x$. Algorithms for finding fixed points of nonexpansive mappings will be studied in Section 6.2.

Moreover, both $P_B \circ P_A$ and $\frac{P_A + P_B}{2}$ are firmly nonexpansive (Definition 2.1.13) and the fact from Exercise 6.2 can be used to obtain weak convergence as in Theorems 6.1.1 and 6.1.4.

6.2 Fixed point approximations

Let again $(\mathcal{H}, d)$ be a Hadamard space and $F: \mathcal{H} \to \mathcal{H}$ be a nonexpansive mapping. We are now interested in the set of fixed points of F, that is,

$$\operatorname{Fix} F := \{x \in \mathcal{H}: x = Fx\} .$$

More precisely we would like to find a point $x \in \operatorname{Fix} F$, but since it is difficult in general, we will content ourselves with a sequence (x_n) converging to a fixed point of F. We present three approximation algorithms each of which generates such an approximating sequence.

Krasnoselski–Mann algorithm

The first method is called the *Krasnoselski–Mann algorithm* and is defined as follows. Given a starting point $x_0 \in \mathcal{H}$, we set

$$x_{n+1} := (1 - \alpha_n) Fx_n + \alpha_n x_n, \qquad n \in \mathbb{N}_0 , \tag{6.2.6}$$

where $\alpha_n \in [0, 1]$ for each $n \in \mathbb{N}_0$. As we shall see in Theorem 6.2.1 below, the sequence obtained by (6.2.6) converges *weakly* to a fixed point.

Theorem 6.2.1. *Let $(\mathcal{H}, d)$ be a Hadamard space and $F: \mathcal{H} \to \mathcal{H}$ be a nonexpansive mapping with* $\operatorname{Fix} F \neq \emptyset$. *Assume that $(\alpha_n) \subset [0, 1]$ satisfying $\sum_n \alpha_n(1-\alpha_n) = \infty$. Then given a starting point $x_0 \in \mathcal{H}$, the sequence (x_n) defined by* (6.2.6) *weakly converges to some $x \in \operatorname{Fix} F$.*

Proof. Step 1: We first establish Fejér monotonicity. Let $z \in \operatorname{Fix} f$. From (1.2.2) we obtain

$$\begin{aligned} d(z, x_{n+1})^2 &\le (1 - \alpha_n) d(Fz, Fx_n)^2 + \alpha_n d(z, x_n)^2 - \alpha_n (1 - \alpha_n) d(Fx_n, x_n)^2 \\ &\le d(z, x_n)^2 - \alpha_n (1 - \alpha_n) d(Fx_n, x_n)^2 , \end{aligned}$$

which implies that the sequence (x_n) is Fejér monotone with respect to $\operatorname{Fix} F$.

Step 2: If one applies Exercise 6.5 to the last inequality, we get

$$\sum_n \alpha_n (1 - \alpha_n) d(Fx_n, x_n)^2 < \infty .$$

By the assumptions on (α_n), we consequently obtain $\liminf d(Fx_n, x_n) = 0$. By the convexity of the metric, we have

$$\begin{aligned} d(Fx_{n+1}, x_{n+1}) &= d(Fx_{n+1}, (1 - \alpha_n) Fx_n + \alpha_n x_n) \\ &\le d(Fx_{n+1}, Fx_n) + d(Fx_n, (1 - \alpha_n) Fx_n + \alpha_n x_n) \\ &\le d(x_{n+1}, x_n) + \alpha_n d(Fx_n, x_n) \\ &= d(Fx_n, x_n) , \end{aligned}$$

and the sequence $(d(Fx_n, x_n))$ hence has a limit. Therefore $\lim d(Fx_n, x_n) = 0$.

Step 3: According to Proposition 3.2.6 it suffices to show that every weak cluster point of (x_n) is a fixed point of F. Let $x \in \mathcal{H}$ be a weak cluster point of (x_n) and let (y_n) be a subsequence of (x_n) such that $y_n \xrightarrow{w} x$. Since $d(y_n, Fy_n) \to 0$, we have

$$\begin{aligned}\limsup_{n\to\infty} d\left(Fx, y_n\right) &\le \limsup_{n\to\infty} \left[d\left(Fx, Fy_n\right) + d\left(Fy_n, y_n\right)\right] \\ &= \limsup_{n\to\infty} d\left(Fx, Fy_n\right) \\ &\le \limsup_{n\to\infty} d\left(x, y_n\right) .\end{aligned}$$

This, by the uniqueness of the weak limit, implies $x = Fx$. Applying Proposition 3.2.6 finishes the proof. □

It is known that the convergence in Theorem 6.2.1 is not strong in general [89]; see also [32, Corollary 5.2].

Halpern algorithm

Next we introduce our second algorithm approximating fixed points, which is attributed to B. Halpern. Given a point $x_0 \in \mathcal{H}$, define

$$x_{n+1} := \left(1 - \alpha_n\right) Fx_n + \alpha_n x_0 , \quad n \in \mathbb{N}_0 . \tag{6.2.7}$$

The strong convergence of the Halpern algorithm is shown in Theorem 6.2.2 below. Notice however that the sequence (6.2.7) is not Fejér monotone.

It is worth mentioning that if F is a linear nonexpansive mapping on a Hilbert space H and we set $\alpha_n := \frac{1}{n+1}$, then the iterations in (6.2.7) become

$$x_n = \frac{1}{n+1} \sum_{k=0}^{n} F^{(k)} x_0,$$

and are usually called the *Cesàro means*, or *ergodic averages*.

Theorem 6.2.2. *Let $(\mathcal{H}, d)$ be a Hadamard space and $F\colon \mathcal{H} \to \mathcal{H}$ be a nonexpansive mapping with* Fix $F \neq \emptyset$. *Assume that a sequence $(\alpha_n) \subset (0, 1)$ satisfies:*
(i) $\alpha_n \to 0$,
(ii) $\sum_n \alpha_n = \infty$,
(iii) $\sum_n |\alpha_n - \alpha_{n-1}| < \infty$.
Given a starting point $x_0 \in \mathcal{H}$, let (x_n) be the sequence defined by (6.2.7). *Then,*

$$\lim_{n\to\infty} x_n = P_{\operatorname{Fix} F}\left(x_0\right) . \tag{6.2.8}$$

Proof. Step 1: First observe that the sequence (x_n) is bounded. By the convexity of the metric we have

$$\begin{aligned} d(x_{n+1}, x_n) &= d\left((1-\alpha_n) Fx_n + \alpha_n x_0, \alpha_{n-1} x_0 + (1-\alpha_{n-1}) Fx_{n-1}\right) \\ &\le d\left((1-\alpha_n) Fx_n + \alpha_n x_0, \alpha_n x_0 + (1-\alpha_n) Fx_{n-1}\right) \\ &\quad + d\left(\alpha_n x_0 + (1-\alpha_n) Fx_{n-1}, \alpha_{n-1} x_0 + (1-\alpha_{n-1}) Fx_{n-1}\right) \\ &\le (1-\alpha_n) d(Fx_n, Fx_{n-1}) + |\alpha_n - \alpha_{n-1}| d(x_0, Fx_{n-1}) \\ &\le (1-\alpha_n) d(x_n, x_{n-1}) + |\alpha_n - \alpha_{n-1}| d(x_0, Fx_{n-1}). \end{aligned}$$

Exercise 6.4 yields that $d(x_{n+1}, x_n) \to 0$. Furthermore,

$$\begin{aligned} d(x_n, Fx_n) &\le d(x_n, x_{n+1}) + d(x_{n+1}, Fx_n) \\ &\le d(x_n, x_{n+1}) + d(\alpha_n x_0 + (1-\alpha_n) Fx_n, Fx_n) \\ &\le d(x_n, x_{n+1}) + \alpha_n d(x_0, Fx_n). \end{aligned}$$

The assumption (i) now gives $d(x_n, Fx_n) \to 0$.

Step 2: Denote $z := P_{\operatorname{Fix} F}(x_0)$. We have

$$\begin{aligned} d(z, x_{n+1})^2 &\le (1-\alpha_n) d(z, Fx_n)^2 + \alpha_n d(z, x_0)^2 - \alpha_n (1-\alpha_n) d(Fx_n, x_0)^2 \\ &\le (1-\alpha_n) d(z, Fx_n)^2 + \alpha_n \left[d(z, x_0)^2 - (1-\alpha_n) d(Fx_n, x_0)^2\right]. \end{aligned}$$

In order to apply the convergence condition from Exercise 6.4 and hence finish the proof, we need to show that

$$\limsup_{n\to\infty} \left[d(z, x_0)^2 - (1-\alpha_n) d(Fx_n, x_0)^2\right] \le 0. \tag{6.2.9}$$

Choose a sequence $\lambda_m \in (0, \infty)$ with $\lim \lambda_m = \infty$. Denote $z_m := R_{\lambda_m} x_0$, where R_λ is the resolvent of F, and observe that this sequence is bounded due to Theorem 4.2.8, because $\operatorname{Fix} F \neq \emptyset$. Then,

$$\begin{aligned} d(x_n, z_m)^2 &\le \frac{\lambda_m}{1+\lambda_m} d(x_n, Fz_m)^2 + \frac{1}{1+\lambda_m} d(x_n, x_0)^2 \\ &\quad - \frac{\lambda_m}{(1+\lambda_m)^2} d(x_0, Fz_m)^2 \\ &\le \frac{\lambda_m}{1+\lambda_m} \left[d(x_n, Fx_n) + d(Fx_n, Fz_m)\right]^2 + \frac{1}{1+\lambda_m} d(x_n, x_0)^2 \\ &\quad - \frac{\lambda_m}{(1+\lambda_m)^2} d(x_0, Fz_m)^2 \\ &\le \frac{\lambda_m}{1+\lambda_m} \left[d(x_n, Fx_n) + d(x_n, z_m)\right]^2 + \frac{1}{1+\lambda_m} d(x_n, x_0)^2 \\ &\quad - \frac{\lambda_m}{(1+\lambda_m)^2} d(x_0, Fz_m)^2. \end{aligned}$$

Since $\lim d(x_n, Fx_n) = 0$, we obtain

$$\frac{\lambda_m}{(1+\lambda_m)^2} d(x_0, Fz_m)^2 \le \frac{1}{1+\lambda_m} \liminf_{n\to\infty} d(x_n, x_0)^2,$$

and after dividing by $\frac{1}{1+\lambda_m}$ and taking the limit $m \to \infty$ it becomes

$$d(x_0, z)^2 \le \liminf_{n\to\infty} d(x_n, x_0)^2.$$

Consequently,

$$d(x_0, z)^2 \le \liminf_{n\to\infty} (1-\alpha_n)\, d(Fx_n, x_0)^2,$$

which already implies (6.2.9). The proof is now complete. □

Yet another fixed point algorithm

We now give one more algorithm for finding a fixed point of a nonexpansive mapping $F\colon \mathcal{H} \to \mathcal{H}$. In Hilbert spaces it coincides with the proximal point algorithm for the maximal monotone operator $A := I - F$, where I is the identity. The algorithm is however probably only of theoretical interest because at each step, we need to find a fixed point of a contraction mapping, which itself requires an approximation algorithm.

Proposition 6.2.3. *Let $F\colon \mathcal{H} \to \mathcal{H}$ be a nonexpansive mapping with at least one fixed point and let $(\lambda_n) \subset (0,\infty)$ be a sequence satisfying $\sum_n \lambda_n^2 = \infty$. Given a point $x_0 \in \mathcal{H}$, put*

$$x_n := R_{\lambda_n} x_{n-1}, \quad n \in \mathbb{N},$$

where R_{λ_n} is the resolvent of F. Then the sequence (x_n) weakly converges to a fixed point of F.

Proof. Let $x \in \mathcal{H}$ be a fixed point of F. Then, for each $n \in \mathbb{N}$, we have

$$d(x_{n-1}, x) \ge d\left(R_{\lambda_n} x_{n-1}, R_{\lambda_n} x\right) = d(x_n, x),$$

which verifies the Fejér monotonicity of (x_n) with respect to $\operatorname{Fix} F$. Put

$$\beta_n := \frac{1}{1+\lambda_n}.$$

Inequality (1.2.2) yields

$$\begin{aligned} d(x, x_n)^2 &\le \beta_n d(x, x_{n-1})^2 + (1-\beta_n) d(x, Fx_n)^2 - \beta_n(1-\beta_n) d(x_{n-1}, Fx_n)^2 \\ &\le \beta_n d(x, x_{n-1})^2 + (1-\beta_n) d(x, x_n)^2 - \beta_n d(x_{n-1}, x_n)^2, \end{aligned}$$

which gives

$$d(x_{n-1}, x_n)^2 \le d(x, x_{n-1})^2 - d(x, x_n)^2,$$

and hence

$$\lambda_n^2 \frac{d(x_{n-1}, x_n)^2}{\lambda_n^2} \le d(x, x_{n-1})^2 - d(x, x_n)^2 . \tag{6.2.10}$$

By the triangle inequality, we have

$$\begin{aligned} d(x_n, x_{n+1}) + d(x_{n+1}, Fx_{n+1}) &= d(x_n, Fx_{n+1}) \\ &\le d(x_n, Fx_n) + d(Fx_n, Fx_{n+1}) \\ &\le d(x_n, Fx_n) + d(x_n, x_{n+1}), \end{aligned}$$

and therefore

$$\frac{d(x_n, x_{n+1})}{\lambda_{n+1}} = d(x_{n+1}, Fx_{n+1}) \le d(x_n, Fx_n) = \frac{d(x_{n-1}, x_n)}{\lambda_n} \tag{6.2.11}$$

Summing up (6.2.10) over $n = 1, \ldots, m$, where $m \in \mathbb{N}$, and using (6.2.11) gives

$$\left(\sum_{n=1}^{m} \lambda_n^2\right) \frac{d(x_{m-1}, x_m)^2}{\lambda_m^2} \le d(x, x_0)^2 - d(x, x_m)^2 .$$

Take the limit $m \to \infty$ to arrive at

$$d(x_m, Fx_m) = \frac{1}{\lambda_m} d(x_{m-1}, x_m) \to 0, \quad \text{as } m \to \infty.$$

Assume now that $z \in \mathcal{H}$ is a weak cluster point of (x_n). Then

$$\begin{aligned} \limsup_{n\to\infty} d(Fz, x_n) &\le \limsup_{n\to\infty} \left[d(Fz, Fx_n) + d(Fx_n, x_n) \right], \\ &\le \limsup_{n\to\infty} d(z, x_n) + 0. \end{aligned}$$

By the uniqueness of the weak limit we get $z = Fz$. Finally apply Proposition 3.2.6 (iii) to conclude that (x_n) weakly converges to a fixed point of F. □

Proposition 6.2.3 with a constant sequence $\lambda_n := \lambda$ for every $n \in \mathbb{N}$ follows by Exercise 6.2.

As already alluded to in Remark 6.1.5 the above algorithms for finding fixed points can be used in convex feasibility problems.

6.3 Proximal point algorithm

Next we are concerned with a minimization problem for a convex lsc function on a Hadamard space. It turns out that the proximal point algorithm (PPA), which is a popular optimization method in Hilbert spaces can be extended into the Hadamard space setting and works equally well. We first introduce a basic form of this algorithm and then turn to its more sophisticated version (so-called splitting PPA) which applies in a rather special albeit very important case.

Basic form of the PPA

The proximal point algorithm is a method for finding a minimizer of a convex lsc function. It can be viewed as a discrete version of the gradient flow semigroup presented in Chapter 5.

Let $(\mathcal{H}, d)$ be a Hadamard space and $f\colon \mathcal{H} \to (-\infty, \infty]$ be a convex lsc function. The *proximal point algorithm* starting at a point $x_0 \in \mathcal{H}$ generates at the n-th step, $n \in \mathbb{N}$, the point

$$x_n := J_{\lambda_n} x_{n-1} = \arg\min_{y\in\mathcal{H}} \left[f(y) + \frac{1}{2\lambda_n} d\left(y, x_{n-1}\right)^2 \right]. \tag{6.3.12}$$

Theorem 6.3.1 (Proximal point algorithm). *Let $(\mathcal{H}, d)$ be a Hadamard space and $f\colon \mathcal{H} \to (-\infty, \infty]$ be a convex lsc function attaining its minimum. Then, for an arbitrary starting point $x_0 \in \mathcal{H}$ and a sequence of positive reals (λ_n) such that $\sum_1^\infty \lambda_n = \infty$, the sequence $(x_n) \subset \mathcal{H}$ defined by (6.3.12) weakly converges to a minimizer of f.*

Proof. The set of minimizers $\operatorname{Min} f$ is by the assumptions nonempty and Theorem 2.2.22 gives that the sequence (x_n) is Fejér monotone with respect to $\operatorname{Min} f$. Without loss of generality we may assume that $f = 0$ on $\operatorname{Min} f$. Choose now $p \in \operatorname{Min} f$ and $k \in \mathbb{N}$ and apply Lemma 2.2.23 with $y := p$ and $x := x_{k-1}$ to obtain

$$\lambda_k f(x_k) \le \frac{1}{2} d(x_{k-1}, p)^2 - \frac{1}{2} d(x_k, p)^2, \tag{6.3.13}$$

From (6.3.13) and from the monotonicity of $(f(x_n))_n$, we have

$$2f\left(x_n\right) \sum_{k=1}^{n} \lambda_k \le 2 \sum_{k=1}^{n} \lambda_k f\left(x_k\right) \le d\left(x_0, p\right)^2 - d\left(x_n, p\right)^2,$$

and thus,

$$f\left(x_n\right) \le \frac{d\left(x_0, p\right)^2}{\sum_{k=1}^{n} \lambda_k}, \tag{6.3.14}$$

for every $n \in \mathbb{N}$. By the assumptions, the right hand side of the last inequality goes to 0 as $n \to \infty$. We thence have that (x_n) is a minimizing sequence, that is, $f(x_n) \to 0$ as $n \to \infty$.

Assume now that a point $z \in \mathcal{H}$ is a weak cluster point of (x_n), that is, there exists a subsequence (x_{n_k}) of (x_n) such that $x_{n_k} \xrightarrow{w} z$. Since f is lsc, and therefore weakly lsc by Lemma 3.2.3, we get $f(z) = 0$ and hence $z \in \operatorname{Min} f$. Applying Proposition 3.2.6 (iii) finally gives $x_n \xrightarrow{w} z$ as $n \to \infty$. This finishes the proof. □

The above proximal point algorithm for a constant sequence $\lambda_n := \lambda$, for every $n \in \mathbb{N}$, follows by Exercise 6.2.

Remark 6.3.2 (Rate of the convergence). As we saw in (6.3.14), for each $n \in \mathbb{N}$, we have

$$f(x_n) - \inf f \leq \frac{K}{\sum_{k=1}^{n} \lambda_k},$$

where K is a positive constant depending on the objective function f and the starting point x_0.

Interestingly, there exist a convex lsc function $f\colon \ell_2 \to (-\infty, \infty]$ attaining its minimum and a point $x_0 \in \ell_2$ such that the PPA starting at x_0 does not converge strongly to a minimizer of f. For more details, see [29, 32, 98]. Consequently, if one wants to obtain strong convergence of a PPA sequence, more assumptions are necessary. For instance, it would be sufficient to require the underlying space to be locally compact. Alternatively, we may require the function f be strongly convex on bounded subsets of its domain, since then every minimizing sequence converges *strongly* to a minimizer by Proposition 2.2.17.

Splitting proximal point algorithm

If the objective function in question has a special form, we can take an advantage of that and improve the proximal point algorithm. We will now study minimization problems for a function $f\colon \mathcal{H} \to (-\infty, \infty]$ of the form

$$f := \sum_{n=1}^{N} f_n, \tag{6.3.15}$$

where $f_n\colon \mathcal{H} \to (-\infty, \infty]$ are all convex and lsc. The main trick here is that instead of applying iteratively the resolvent

$$J_\lambda x := \operatorname*{arg\,min}_{y \in \mathcal{H}} \left[f(y) + \frac{1}{2\lambda} d(x, y)^2 \right]$$

of the function f, we apply the resolvents

$$J_\lambda^n x := \operatorname*{arg\,min}_{y \in \mathcal{H}} \left[f_n(y) + \frac{1}{2\lambda} d(x, y)^2 \right] \tag{6.3.16}$$

of its components f_n. Such a method is called the *splitting proximal point algorithm*, and abbreviated SPPA.

There are essentially two ways of applying the resolvents (6.3.16) in the SPPA. We either fix an order of the components (that is, a permutation of the numbers $1, \dots, N$, which without loss of generality may be the identity permutation) and at each cycle we will apply the resolvents sequentially in this fixed order, or alternatively, we will at each step pick a number $r \in \{1, \dots, N\}$ at random and apply the resolvent of f_r.

Theorems 6.3.7 and 6.3.13 below say that in either case, we get a sequence converging to a minimizer of f.

Note that in order to obtain the convergence of the SPPA, the underlying space is assumed to be locally compact.

Question 6.3.3. Is it possible to obtain *weak* convergence in Theorems 6.3.7 and 6.3.13 below, provided we drop the assumption the space be locally compact?

Prototypical instances of (6.3.15) are in Examples 2.2.5 and 2.2.6.

Example 6.3.4 (Computing medians and means). Given a finite number of points $a_1, \ldots, a_N \in \mathcal{H}$ and $(w_1, \ldots, w_N) \in \Delta_{N-1}$, we consider the function

$$f(x) := \sum_{n=1}^{N} w_n d\left(x, a_n\right)^p, \quad x \in \mathcal{H}, \tag{6.3.17}$$

where $p \in [1, \infty)$. In this case, computing the resolvent J_λ^n is actually a one-dimensional problem. In other words, employing the splitting PPA reduces the computation of a minimizer of f to a sequence of one-dimensional problems. We are in particular interested in computing medians and means, that is, the cases $p = 1$ and $p = 2$, respectively. This will be greatly exploited in Section 8.3.

Interestingly, a mean can be also computed via the law of large numbers; see Algorithm 7.2.3.

Example 6.3.5 (Constrained minimization). Let $f \colon \mathcal{H} \to (-\infty, \infty]$ be convex lsc and $C \in \mathcal{H}$ be closed convex. To solve the *constrained* minimization problem

$$\operatorname*{arg\,min}_{x \in C} f(x),$$

one can write it in an equivalent form

$$\operatorname*{arg\,min}_{x \in \mathcal{H}} g(x),$$

where $g := f + \iota_C$ and apply the splitting PPA. Recall that the resolvent of ι_C is P_C, the metric projection onto C.

Cyclic order version of the SPPA

We will now prove the first main result, namely, that the proximal point algorithm with the cyclic order of applying the marginal resolvents (6.3.16) gives a sequence which converges to a minimizer. Let us first precisely define the procedure.

Consider a function f of the form (6.3.15). Let (λ_k) be a sequence of positive reals satisfying

$$\sum_{k=0}^{\infty} \lambda_k = \infty, \quad \text{and} \quad \sum_{k=0}^{\infty} \lambda_k^2 < \infty. \tag{6.3.18}$$

Let $x_0 \in \mathcal{H}$ be an arbitrary starting point. For each $k \in \mathbb{N}_0$ we set

$$\begin{aligned} x_{kN+1} &:= J_{\lambda_k}^1 \left(x_{kN}\right) , \\ x_{kN+2} &:= J_{\lambda_k}^2 \left(x_{kN+1}\right) , \\ &\vdots \\ x_{kN+N} &:= J_{\lambda_k}^N \left(x_{kN+N-1}\right) , \end{aligned} \tag{6.3.19}$$

where the resolvents are defined by (6.3.16) above.

Note that the step size parameter λ_k is constant throughout each cycle. The convergence of the above algorithm is assured by the following theorem. The assumption (6.3.20) will be commented on in Remark 6.3.14.

Example 6.3.6 (Alternating projections revisited). Let $(\mathcal{H}, d)$ be a Hadamard space and $A, B \subset \mathcal{H}$ be closed convex with $A \cap B \neq \emptyset$. Then the cyclic version of the SPPA applied to the function

$$f := \iota_A + \iota_B$$

generates the same sequence as the alternating projection method from Section 6.1.

Theorem 6.3.7 (SPPA with cyclic order). *Let $(\mathcal{H}, d)$ be a locally compact Hadamard space and $f : \mathcal{H} \to (-\infty, \infty]$ be of the form* (6.3.15) *with* $\operatorname{Min} f \neq \emptyset$. *Given a starting point $x_0 \in \mathcal{H}$, let (x_j) be the sequence defined in* (6.3.19). *Assume there exists $L > 0$ such that*

$$f_n \left(x_{kN}\right) - f_n \left(x_{kN+n}\right) \leq L d \left(x_{kN}, x_{kN+n}\right) , \tag{6.3.20a}$$

$$f_n \left(x_{kN+n-1}\right) - f_n \left(x_{kN+n}\right) \leq L d \left(x_{kN+n-1}, x_{kN+n}\right), \tag{6.3.20b}$$

for each $k \in \mathbb{N}_0$ and $n = 1, \ldots, N$. Then (x_j) converges to a minimizer of f.

Proof. We divide the proof into two steps. Step 1: We claim that

$$d \left(x_{kN+N}, y\right)^2 \leq d \left(x_{kN}, y\right)^2 - 2\lambda_k \left[f \left(x_{kN}\right) - f(y)\right] + 2\lambda_k^2 L^2 N(N+1), \tag{6.3.21}$$

for every $y \in \mathcal{H}$. Indeed, apply Lemma 2.2.23 to obtain

$$d \left(x_{kN+n}, y\right)^2 \leq d \left(x_{kN+n-1}, y\right)^2 - 2\lambda_k \left[f_n \left(x_{kN+n}\right) - f_n(y)\right],$$

for every $y \in \mathcal{H}$ and $n = 1, \ldots, N$. By summing up we obtain

$$\begin{aligned} d \left(x_{kN+N}, y\right)^2 &\leq d \left(x_{kN}, y\right)^2 - 2\lambda_k \sum_{n=1}^{N} \left[f_n \left(x_{kN+n}\right) - f_n(y)\right] , \\ &= d \left(x_{kN}, y\right)^2 - 2\lambda_k \left[f \left(x_{kN}\right) - f(y)\right] \\ &\quad + 2\lambda_k \sum_{n=1}^{N} \left[f_n \left(x_{kN}\right) - f_n \left(x_{kN+n}\right)\right] . \end{aligned}$$

By assumption (6.3.20a), we have

$$f_n(x_{kN}) - f_n(x_{kN+n}) \leq Ld(x_{kN}, x_{kN+n}),$$

where the right hand side can be further estimated as

$$d(x_{kN}, x_{kN+n}) \leq d(x_{kN}, x_{kN+1}) + \cdots + d(x_{kN+n-1}, x_{kN+n}).$$

By the definition of the algorithm we have

$$f_n(x_{kN+n}) + \frac{1}{2\lambda_k} d(x_{kN+n-1}, x_{kN+n})^2 \leq f_n(x_{kN+n-1}),$$

for each $n = 1, \ldots, N$, which then gives

$$\begin{aligned} d(x_{kN+n-1}, x_{kN+n}) &\leq 2\lambda_k \frac{f_n(x_{kN+n-1}) - f_n(x_{kN+n})}{d(x_{kN+n-1}, x_{kN+n})} \\ &\leq 2\lambda_k L, \end{aligned} \tag{6.3.22}$$

where we employed assumption (6.3.20b). Hence,

$$f_n(x_{kN}) - f_n(x_{kN+n}) \leq 2\lambda_k L^2 n,$$

and finally,

$$d(x_{kN+N}, y)^2 \leq d(x_{kN}, y)^2 - 2\lambda_k [f(x_{kN}) - f(y)] + 2\lambda_k^2 L^2 N(N+1),$$

which finishes the proof of (6.3.21).

Step 2: Let now $z \in \operatorname{Min} f$, and apply (6.3.21) with $y := z$. Then

$$d(x_{kN+N}, z)^2 \leq d(x_{kN}, z)^2 - 2\lambda_k [f(x_{kN}) - f(z)] + 2\lambda_k^2 L^2 N(N+1),$$

which according to Exercise 6.5 implies that the sequence

$$(d(x_{kN}, z))_{k \in \mathbb{N}_0}$$

converges, (and in particular, the sequence (x_{kN}) is bounded), and

$$\sum_{k=0}^{\infty} \lambda_k [f(x_{kN}) - f(z)] < \infty. \tag{6.3.23}$$

From (6.3.23) we immediately obtain that there exists a subsequence $(x_{k_l N})$ of (x_{kN}) for which

$$f(x_{k_l N}) \to f(z), \quad \text{as } l \to \infty.$$

Since the sequence $(x_{k_l N})$ is bounded, it has a subsequence which converges to a point $\widehat{z} \in \mathcal{H}$. By the lower semicontinuity of f we obtain $\widehat{z} \in \operatorname{Min} f$. Then we know that

$$(d(x_{kN}, \widehat{z}))_{k \in \mathbb{N}_0}$$

converges, and also that it converges to 0, since a subsequence of (x_{kN}) converges to $\widehat{z}$.

By virtue of (6.3.22), we obtain

$$\lim_{k\to\infty} x_{kN+n} = \widehat{z},$$

for every $n = 1, \ldots, N$. Hence the whole sequence (x_j) converges to $\widehat{z}$ and the proof is complete. □

Random order variant of the SPPA

Instead of applying the marginal resolvents in a cyclic order, one can at each step select a number from $\{1, \ldots, N\}$ at random and apply the corresponding resolvent. Using some tools from probability theory, namely a refined form of the classical supermartingale convergence theorem, we will prove that the resulting SPPA sequence converges to a minimizer of the function f.

Let us first recall some terminology and notation. Let $(\Omega, \mathcal{F})$ a measurable space. A sequence $(\mathcal{F}_k)_{k\in\mathbb{N}_0}$ of σ-algebras on Ω is called a *filtration* if $\mathcal{F}_0 \subset \mathcal{F}_1 \subset \cdots \subset \mathcal{F}$. Then $(\Omega, \mathcal{F}, (\mathcal{F}_k))$ is a *filtered measurable space*. A filtered measurable space with a probability measure is called a *filtered probability space*. We also denote

$$\mathcal{F}_\infty := \sigma\Bigg(\bigcup_{k\in\mathbb{N}_0} \mathcal{F}_k\Bigg),$$

and observe that $\mathcal{F}_\infty$ is a sub-σ-algebra of $\mathcal{F}$. Assume that

$$\Big(\Omega, \mathcal{F}, (\mathcal{F}_k)_{k\in\mathbb{N}_0}, \mathbb{P}\Big)$$

is now a fixed filtered probability space.

Definition 6.3.8 (Nonnegative supermartingale). A nonnegative stochastic process (X_k), where $k \in \mathbb{N}_0$, is a *supermartingale* if
(i) X_k is $\mathcal{F}_k$-measurable,
(ii) $\mathbb{E}_{\mathcal{F}_k} X_{k+1} \leq X_k$,
for each $k \in \mathbb{N}_0$. Note that if we have *equality* in (ii), the process is called a *martingale*.

The following theorem is a fundamental result in martingale theory.

Theorem 6.3.9 (Supermartingale convergence theorem). *Let (X_k) be a nonnegative supermartingale. Then there exists a finite nonnegative random variable X which is $\mathcal{F}_\infty$-measurable and (X_k) converges to X almost surely.*

Proof. See [204, Corollary 11.7]. □

We shall need a slightly improved form of the above theorem. In order to make the proof more transparent, we state its intermediate version as a lemma.

Lemma 6.3.10. *Let (X_k) and (W_k) be a nonnegative stochastic processes such that:*
(i) X_k *and* W_k *are* $\mathcal{F}_k$*-measurable for each* $k \in \mathbb{N}_0$.
(ii) $\mathbb{E}_{\mathcal{F}_k} X_{k+1} \le X_k + W_k$, *for each* $k \in \mathbb{N}_0$.
(iii) $\sum_k W_k < \infty$.
Then (X_k) converges almost surely to a finite nonnegative random variable.

Proof. Denote

$$U_0 := X_0 , \quad U_k := X_k - \sum_{i=0}^{k-1} W_i , \quad k \in \mathbb{N},$$

and observe

$$\mathbb{E}_{\mathcal{F}_k} U_{k+1} = \mathbb{E}_{\mathcal{F}_k} X_{k+1} - \sum_{i=0}^{k} W_i \le X_k + W_k - \sum_{i=0}^{k} W_i = U_k,$$

for each $k \in \mathbb{N}_0$. Given $\alpha \in (0, \infty)$, we define a random variable $\nu_\alpha := \inf\{n \in \mathbb{N}_0 : \sum_{i=0}^{n} W_i > \alpha\}$. Then

$$\left((\alpha + U_k) \, \chi_{\{\nu_\alpha > k\}} \right)_{k \in \mathbb{N}_0}$$

is a nonnegative supermartingale and hence converges almost surely by Theorem 6.3.9. Hence (U_k) converges almost surely on

$$\{\nu_\alpha = \infty\} = \left\{ \sum_{k=0}^{\infty} W_k \le \alpha \right\} .$$

Since α was arbitrary, it follows that (U_k) converges almost surely on Ω. Consequently, the process (X_k) converges almost surely on Ω to a finite random variable. □

We can now state the promised refinement of the supermartingale convergence theorem.[1]

Theorem 6.3.11. *Assume (Y_k), (Z_k) and (W_k) are sequences of nonnegative real-valued random variables defined on Ω and assume that:*
(i) Y_k, Z_k, W_k *are* $\mathcal{F}_k$*-measurable for each* $k \in \mathbb{N}_0$.
(ii) $\mathbb{E}_{\mathcal{F}_k} Y_{k+1} \le Y_k - Z_k + W_k$, *for each* $k \in \mathbb{N}_0$.
(iii) $\sum_k W_k < \infty$.
Then the sequence (Y_k) converges to a finite random variable Y almost surely, and $\sum_k Z_k < \infty$ almost surely.

Proof. Lemma 6.3.10, applied with $X_k := Y_k$ yields the claim on the convergence of (Y_k). To show that $\sum_k Z_k < \infty$ almost surely, we put

$$V_0 := Y_0 , \quad V_k := Y_k + \sum_{i=0}^{k-1} Z_i , \quad k \in \mathbb{N},$$

1 Its deterministic counterpart is in Exercise 6.5 and was used in the proof of Theorem 6.3.7.

and observe that

$$\mathbb{E}_{\mathcal{F}_k} V_{k+1} = \mathbb{E}_{\mathcal{F}_k} Y_{k+1} + \sum_{i=0}^{k} Z_i \le Y_k - Z_k + W_k + \sum_{i=0}^{k} Z_i = V_k + W_k.$$

Apply Lemma 6.3.10 again, this time with $X_k := V_k$, and conclude that (V_k) converges almost surely to a finite random variable V. Then $\sum_{k=0}^{\infty} Z_k = V - Y$ and the proof is complete. □

For the purposes of the randomized SPPA, consider the probability space $\Omega := \{1, \ldots, N\}^{\mathbb{N}_0}$ equipped with the product of the uniform probability measure on $\{1, \ldots, N\}$ and let (r_k) be the sequence of random variables

$$r_k(\omega) := \omega_k, \quad \omega = (\omega_1, \omega_2, \ldots) \in \Omega.$$

Let $f\colon \mathcal{H} \to (-\infty, \infty]$ be a function of the form (6.3.15) and let (λ_k) be a sequence of positive reals satisfying

$$\sum_{k=0}^{\infty} \lambda_k = \infty, \quad \text{and} \quad \sum_{k=0}^{\infty} \lambda_k^2 < \infty. \tag{6.3.24}$$

Given a starting point $x_0 \in \mathcal{H}$, we put

$$x_{k+1} := J_{\lambda_k}^{r_k} x_k, \quad k \in \mathbb{N}_0. \tag{6.3.25}$$

Define the natural filtration $\mathcal{F}_k := \sigma(x_0, \ldots, x_k)$ on Ω, and finally, denote x_{k+1}^n the result of the iteration (6.3.25) with x_k if $r_k(\omega) = n$.

Having Theorem 6.3.11 at hand, we prove the following lemma. Again, the assumption (6.3.26) below will be commented on in Remark 6.3.14.

Lemma 6.3.12. *Let $(\mathcal{H}, d)$ be a locally compact Hadamard space and f be of the form (6.3.15). Given a starting point $x_0 \in \mathcal{H}$, let (x_k) be the sequence defined in (6.3.25). Assume there exists $L > 0$ such that*

$$f_n(x_k) - f_n(x_{k+1}^n) \le L d(x_k, x_{k+1}^n), \tag{6.3.26}$$

for every $k \in \mathbb{N}_0$ and $n = 1, \ldots, N$. Then we have

$$\mathbb{E}_{\mathcal{F}_k} d(x_{k+1}, y)^2 \le d(x_k, y)^2 - \frac{2\lambda_k}{N}[f(x_k) - f(y)] + 4\lambda_k^2 L^2,$$

almost surely, for every $y \in \mathcal{H}$.

Proof. By Lemma 2.2.23 we have

$$d(x_{k+1}, y)^2 \le d(x_k, y)^2 - 2\lambda_k \left[f_{r_k}(x_{k+1}) - f_{r_k}(y)\right].$$

Taking the conditional expectation with respect to $\mathcal{F}_k$ gives

$$\mathbb{E}_{\mathcal{F}_k} d(x_{k+1}, y)^2 \le d(x_k, y)^2 - 2\lambda_k \mathbb{E}_{\mathcal{F}_k}\left[f_{r_k}(x_{k+1}) - f_{r_k}(y)\right].$$

If we denote x_{k+1}^n the result of the iteration with x_k when $r_k = n$, we get

$$\begin{aligned}\mathbb{E}_{\mathcal{F}_k} d(x_{k+1}, y)^2 &\le d(x_k, y)^2 - \frac{2\lambda_k}{N} \sum_{n=1}^{N} \left[f_n(x_{k+1}^n) - f_n(y)\right] \\ &= d(x_k, y)^2 - \frac{2\lambda_k}{N} \left[f(x_k) - f(y)\right] \\ &\quad + \frac{2\lambda_k}{N} \sum_{n=1}^{N} \left[f_n(x_k) - f_n(x_{k+1}^n)\right].\end{aligned}$$

By assumption (6.3.26) we have

$$\sum_{n=1}^{N} \left[f_n(x_k) - f_n(x_{k+1}^n)\right] \le L \sum_{n=1}^{N} d(x_k, x_{k+1}^n) \le 2L^2\lambda_k N,$$

since

$$d(x_k, x_{k+1}^n) \le 2\lambda_k \frac{f_n(x_k) - f_n(x_{k+1}^n)}{d(x_k, x_{k+1}^n)} \le 2\lambda_k L.$$

We hence finally obtain

$$\mathbb{E}_{\mathcal{F}_k} d(x_{k+1}, y)^2 \le d(x_k, y)^2 - \frac{2\lambda_k}{N} \left[f(x_k) - f(y)\right] + 4\lambda_k^2 L^2,$$

which finishes the proof. □

We now prove the convergence of the random order version of the SPPA.

Theorem 6.3.13 (SPPA with random order). *Let $(\mathcal{H}, d)$ be a locally compact Hadamard space and f be of the form (6.3.15) with $\operatorname{Min} f \neq \emptyset$. Assume that the Lipschitz condition (6.3.26) holds true. Then, given a starting point $x_0 \in \mathcal{H}$, the sequence (x_k) defined in (6.3.25) converges to a minimizer of f almost surely.*

Proof. Since $\operatorname{Min}(f)$ is a locally compact Hadamard space, its closed balls are compact by the Hopf–Rinow theorem [51, p. 35] and consequently it is separable. We can thus choose a countable dense subset $\{v_i\}$ of $\operatorname{Min}(f)$. For each $i \in \mathbb{N}$ apply Lemma 6.3.12 with $y := v_i$ to obtain

$$\mathbb{E}_{\mathcal{F}_k} d(x_{k+1}(\omega), v_i)^2 \le d(x_k(\omega), v_i)^2 - \frac{2\lambda_k}{N} \left[f(x_k(\omega)) - f(v_i)\right] + 4\lambda_k^2 L^2.$$

for every ω from a full measure set $\Omega_{v_i} \subset \Omega$. Theorem 6.3.11 immediately gives that $d(v_i, x_k(\omega))$ converges, and

$$\sum_{k=0}^{\infty} \lambda_k \left[f(x_k(\omega)) - \inf f\right] < \infty,$$

for all $\omega \in \Omega_{v_i}$, where Ω_{v_i} is a set of full measure. Next denote

$$\Omega_\infty := \bigcap_{i \in \mathbb{N}} \Omega_{v_i},$$

which is by countable subadditivity again a set of full measure. The last inequality yields that for $\omega \in \Omega_\infty$, we have $\liminf_{k\to\infty} f(x_k(\omega)) = \inf f$, and since $(x_k(\omega))$ is bounded, it has a cluster point $x(\omega) \in \mathcal{H}$. By the lower semicontinuity of f we know that $x(\omega) \in \operatorname{Min} f$.

For each $\varepsilon > 0$ there exists $v_{i(\varepsilon)} \in (v_i)$ such that $d(x(\omega), v_{i(\varepsilon)}) < \varepsilon$. Because the sequence $d(x_k(\omega), v_{i(\varepsilon)})$ converges and $x(\omega)$ is a cluster point of $x_k(\omega)$, we have

$$\lim_{k\to\infty} d\left(x_k(\omega), v_{i(\varepsilon)}\right) < \varepsilon.$$

This yields $x_k(\omega) \to x(\omega)$. We obtain that x_k converges to a minimizer almost surely. This finishes the proof. □

Remark 6.3.14. The assumptions (6.3.20) in Theorem 6.3.7, and (6.3.26) in Theorem 6.3.13 are satisfied, for instance, if

(i) the functions f_n are Lipschitz on $\mathcal{H}$ with constant L, or
(ii) the SPPA sequence (6.3.19), or (6.3.25) respectively, is bounded and the functions f_n are Lipschitz on this bounded set with constant L.

To summarize the most important cases of Theorems 6.3.7 and 6.3.13, we state the following corollary.

Corollary 6.3.15. *Let $(\mathcal{H}, d)$ be a locally compact Hadamard space and f be of the form* (6.3.15). *Assume that (at least) one of the following conditions is satisfied:*

(i) *the functions f_n are Lipschitz on $\mathcal{H}$ with constant L and* $\operatorname{Min} f \neq \emptyset$, *or*
(ii) *the function f is of the form* (6.3.17).

Then:

(i) *The sequence defined in* (6.3.19) *converges to a minimizer of f.*
(ii) *The sequence defined in* (6.3.25) *converges to a minimizer of f almost surely.*

Proof. The proof follows immediately from Remark 6.3.14 and the fact that if the function f is of the form (6.3.17), the SPPA sequences (6.3.19) and (6.3.25) are contained in the bounded set $\overline{\operatorname{co}}\{x_0, a_1, \ldots, a_N\}$. □

Exercises

Exercise 6.1. Give the details in the proof of Theorem 6.1.4.

Exercise 6.2. Let $(\mathcal{H}, d)$ be a Hadamard space and $F\colon \mathcal{H} \to \mathcal{H}$ be firmly nonexpansive with $\operatorname{Fix} F \neq \emptyset$. Given $x_0 \in \mathcal{H}$, put

$$x_n := F x_{n-1}, \quad n \in \mathbb{N}.$$

Show that the sequence (x_n) weakly converges to a fixed point of F.
Hint. The sequence is Fejér monotone with respect to $\operatorname{Fix} F$. Show that $d(x_n, x_{n+1}) \to 0$ and all cluster points of (x_n) lie in $\operatorname{Fix} F$. □

Exercise 6.3. Let $F: \mathcal{H} \to \mathcal{H}$ be nonexpansive and $\alpha \in (0, 1)$. Is then $(1 - \alpha)F + \alpha I$, where $I: \mathcal{H} \to \mathcal{H}$ stands for the identity mapping, a *firmly* nonexpansive mapping?

Exercise 6.4. Let (a_n) and (c_n) be sequences of nonnegative numbers such that $\sum c_n < \infty$ and $(b_n) \subset \mathbb{R}$ be a sequence with $\limsup b_n \le 0$. Let $(t_n) \subset [0, 1]$ be a sequence with $\sum t_n = \infty$. Assume that

$$a_{n+1} \le (1 - t_n)\, a_n + t_n b_n + c_n,$$

for every $n \in \mathbb{N}$. Show that $\lim a_n = 0$.
Hint. Use recursion to estimate a_{n+m} from above. If in trouble, see [10, Lemma 2.3]. □

The following is a deterministic version of Theorem 6.3.11.

Exercise 6.5. Let (a_k), (b_k) and (c_k) be sequences of nonnegative real numbers. Assume that

$$a_{k+1} \le a_k - b_k + c_k, \tag{6.3.27}$$

for each $k \in \mathbb{N}$ and $\sum_{k=1}^{\infty} c_k < \infty$. Show that the sequence (a_k) converges and $\sum_{k=1}^{\infty} b_k < \infty$.
Hint. Fix $l \in \mathbb{N}$. Sum (6.3.27) over $k \ge l$ and take $\limsup_{k\to\infty}$ to obtain

$$\limsup_{k\to\infty} a_k \le a_l + \sum_{k=l}^{\infty} c_k.$$

Taking $\liminf_{l\to\infty}$ yields $\limsup_{k\to\infty} a_k \le \liminf_{l\to\infty} a_l$, and hence (a_k) converges. Now fix $n \in \mathbb{N}$ and sum (6.3.27) from $k = 1$ to $k = n$,

$$\sum_{k=1}^{n} b_k \le a_1 + \sum_{k=1}^{n} c_k - a_{n+1}.$$

Since the last inequality holds for any $n \in \mathbb{N}$, we get $\sum_{k=1}^{\infty} b_k < \infty$. □

Bibliographical remarks

Convex feasibility problems are a classical topic in optimization. The alternating projection method appeared in [22] as a modification of its earlier Hilbert space variant from [28]. The proofs of classical results due to J. von Neumann and L. Bregman can be found in [32, Section 3]. A simple geometric proof of von Neumann's theorem is given in the paper [118] by E. Kopecká and S. Reich. For the construction from Example 6.1.2, see [100, 138]. The Hilbert ball case was studied by S. Reich [172]. For recent information regarding algorithm (6.1.1) in Hilbert space, see the paper [120] by E. Kopecká and S. Reich. It is known that Question 6.1.3 has the positive answer if $\operatorname{codim} A = 1$, but is otherwise largely open [28].

The averaged projection method in Hilbert spaces was shown to weakly converge by A. Auslender [14]. A counterexample to strong convergence was provided in [32] as late as 2004. To the best of our knowledge, Theorem 6.1.4 appears nowhere in the literature.

Theorem 6.2.1 copies its Hilbert space version from [31, Theorem 5.14]. This approximation algorithm has been used extensively since its discovery by M. Krasnoselski [123] and W. Mann [136]. For further analyses, see the papers [129] and [117]. The Krasnoselski–Mann algorithm was studied also in [45, 167].

Theorem 6.2.2 comes from [181] and extends a Hilbert space result of R. Wittmann [205]. We give a simpler proof, which avoids Banach limits and exploits only the geometry of nonpositive curvature. The conditions (i)–(iii) in Theorem 6.2.2 on the parameters λ_n are discussed in detail in [99, 205]. For a deeper study of the Halpern algorithm using *proof mining*, see the papers [115, 116] by U. Kohlenbach and L. Leustean. An authoritative reference on proof mining is the monograph [114] by U. Kohlenbach. We would like to mention a recent paper of S. Reich and L. Shemen which is devoted to the Halpern algorithm in the Hilbert ball [174].

Proposition 6.2.3 appeared in [21]. In a Hilbert space H it corresponds to the PPA for the maximal monotone operator $I - F$, where $F\colon H \to H$; see [31, Theorem 23.41]. For earlier results in nonlinear spaces, see also [52, Theorem 2.6], [173, Corollary 7.10], [187, Theorem 4.7], and [11, Theorem 6.4]. Related results in Banach spaces are contained in the paper [151] by O. Nevanlinna and S. Reich.

The proximal point algorithm originated in the celebrated papers [47, 137, 180]. O. Güler was the first to give a counterexample to strong convergence [98] and newer counterexamples appeared later in [29, 32]. Recent and unifying treatment of the PPA and be found in [31]. The proximal point algorithm in the context of Riemannian manifold (of nonpositive sectional curvature) was studied in [36, 85, 130, 160]. The proof of Theorem 6.3.1 comes from [19]. For another approach to the PPA in metric spaces, see [211] by A. Zaslavski.

The splitting PPA appeared in [16], based on its Euclidean space predecessor due to D. Bertsekas [40]. The cyclic order version of the SPPA was later worked out also in Alexandrov spaces by S. Ohta and M. Pálfia [155]. In the Hadamard space setting, S. Banert studied the SPPA for two functions in [27]. Infinite products of resolvents of accretive operators were studied in [177] by S. Reich and A. Zaslavski. We note that splitting methods go back to P.-L. Lions and B. Mercier [133] and G. Passty [163]. For a nice survey, see [66].

Martingale theory is explained in a number of books on probability. We used [204] by D. Williams. Theorem 6.3.9 is [204, Corollary 11.7]. Lemma was in a more general form proved in [44, Theorem 3.3.6]; see also [152, Exercise II-4]. Theorem 6.3.11 is is stated in [41, Proposition 4.2] by D. Bertsekas and J. Tsitsiklis. We reproduce its proof from unpublished notes [210] by H. Yu.

The statement in Exercise 6.2 appeared in [11, Theorem 6.4]. The claim in Exercise 6.5 comes from [41, Lemma 3.4].

There exist many interesting results in optimization in Hadamard manifolds which have so far no analogues in general Hadamard spaces. Let us mention for instance the paper [131].

I would like to thank Brailey Sims for helpful discussions about fixed point approximations in Hadamard spaces.

7 Probabilistic tools in Hadamard spaces

We will now study random variables with values in a Hadamard space. Probability theory with such mappings was developed by K.-T. Sturm and includes also martingales and Markov processes, which are necessary for a stochastic approach to harmonic maps with values in a Hadamard space. We will however mention only rudiments of this rich probability theory that are needed for our purposes.

As applications, we reprove the Jensen inequality, employ the law of large numbers as an algorithm for computing the Fréchet mean and prove a Hadamard space variant of the Wolfowitz theorem, which consequently gives a consensus algorithm.

7.1 Random variables and expectations

We start by introducing basic objects. Let $(\Omega, \mathcal{F}, \mathbb{P})$ be a probability space and $(\mathcal{H}, d)$ be a Hadamard space. We say that a mapping $Y\colon \Omega \to \mathcal{H}$ is a *random variable* if it is $\mathcal{F}$-measurable, that is, if

$$Y^{-1}(A) \in \mathcal{F},$$

for each Borel set $A \subset \mathcal{H}$. A random variable $Y\colon \Omega \to \mathcal{H}$ defines a probability measure $Y_*\mathbb{P} \in \mathcal{P}(\mathcal{H})$ by push-forward of $\mathbb{P}$ under Y, that is,

$$Y_*\mathbb{P}(A) := \mathbb{P}\left(Y^{-1}(A)\right),$$

where $A \subset \mathcal{H}$ is Borel measurable. The measure $Y_*\mathbb{P}$ is called the *distribution* of Y and denoted by $\mathbb{P}_Y$. We hence have a probability measure on a Hadamard space and can use the results of Section 2.3. Also note that $Y \in \mathcal{L}^p(\Omega, \mathcal{H})$ if and only if $\mathbb{P}_Y \in \mathcal{P}^p(\mathcal{H})$, where the spaces $\mathcal{L}^p(\Omega, \mathcal{H})$ were introduced on page 18. Define the *variance* of Y by

$$\operatorname{var} Y := \operatorname{var} \mathbb{P}_Y = \inf_{x \in \mathcal{H}} \mathbb{E} d\,(x, Y(\omega))^2 \,. \tag{7.1.1}$$

Obviously, $\operatorname{var} Y < \infty$ if and only if $Y \in \mathcal{L}^2(\Omega, \mathcal{H})$.

Definition 7.1.1 (Expectation). The *expectation* of a random variable $Y \in \mathcal{L}^1(\Omega, \mathcal{H})$ is defined as

$$\mathbb{E}Y := b\left(\mathbb{P}_Y\right) \,,$$

that is, as the barycenter of the distribution $\mathbb{P}_Y$. Recall that a barycenter was defined in Theorem 2.3.1.

If $Y\colon \Omega \to [-\infty, \infty]$ is a random variable, then our definition of the expectation coincides with the usual one; see Example 2.3.2. We can hence use the symbol $\mathbb{E}$ in both cases.

If we identify $\mathcal{H}$ with the constant maps in $\mathcal{L}^2(\Omega, \mathcal{H})$, then the map $\mathbb{E}\colon \mathcal{L}^2(\Omega, \mathcal{H}) \to \mathcal{H}$ can be regarded as the metric projection onto the closed convex set of constant maps. Inequality (2.1.1) then reads

$$\mathbb{E}d(z, Y)^2 \geq d(z, \mathbb{E}Y)^2 + \mathbb{E}d(Y, \mathbb{E}Y)^2 , \tag{7.1.2}$$

or, equivalently,

$$\mathbb{E}d(z, Y)^2 \geq d(z, \mathbb{E}Y)^2 + \operatorname{var} Y ,$$

for every $z \in \mathcal{H}$ and $Y \in \mathcal{L}^2(\Omega, \mathcal{H})$. It is exactly the variance inequality (2.3.14) for the measure $\mathbb{P}_Y \in \mathcal{P}^2(\mathcal{H})$.

Proposition 7.1.2. *If $X, Y \in \mathcal{L}^1(\Omega, \mathcal{H})$, then $d(\mathbb{E}X, \mathbb{E}Y) \leq \mathbb{E}d(X, Y)$. In particular, if a sequence $X_n \in \mathcal{L}^1(\Omega, \mathcal{H})$ converges to some $X \in \mathcal{L}^1(\Omega, \mathcal{H})$ in $\mathcal{L}^1(\Omega, \mathcal{H})$, then $\mathbb{E}X_n \to \mathbb{E}X$.*

Proof. Inequality (2.3.14) gives

$$d(\mathbb{E}Y, \mathbb{E}X)^2 \leq \int_\Omega \left[d(\mathbb{E}Y, X(\omega))^2 - d(\mathbb{E}X, X(\omega))^2 \right] \mathrm{d}\mathbb{P}(\omega),$$

as well as,

$$d(\mathbb{E}X, \mathbb{E}Y)^2 \leq \int_\Omega \left[d(\mathbb{E}X, Y(\omega))^2 - d(\mathbb{E}Y, Y(\omega))^2 \right] \mathrm{d}\mathbb{P}(\omega) .$$

Apply Corollary 1.2.5 to the points $X(\omega), Y(\omega), \mathbb{E}X$, and $\mathbb{E}Y$ to obtain

$$d(\mathbb{E}Y, X)^2 - d(\mathbb{E}X, X)^2 + d(Y, \mathbb{E}X)^2 - d(\mathbb{E}Y, Y)^2 \leq 2d(\mathbb{E}X, \mathbb{E}Y)\, d(X, Y) ,$$

almost surely on Ω. Take expectations in the last inequality and sum up all the inequalities above to get

$$2d(\mathbb{E}X, \mathbb{E}Y)^2 \leq 2d(\mathbb{E}X, \mathbb{E}Y)\, \mathbb{E}d(X, Y) ,$$

or,

$$d(\mathbb{E}X, \mathbb{E}Y) \leq \mathbb{E}d(X, Y) , \tag{7.1.3}$$

which is the first claim. The second part is a direct consequence. □

Before moving on to the next section, we recall a few elementary definitions, which readily extend for random variables with values in a Hadamard space $(\mathcal{H}, d)$. Random variables $(Y_n)_{n \in \mathbb{N}}$ are *independent* if for every finite set of indices $i_1, \ldots, i_N \in \mathbb{N}$ and Borel measurable sets $A_{i_1}, \ldots, A_{i_N} \subset \mathcal{H}$, the events

$$\left\{ Y_{i_1} \in A_{i_1} \right\}, \ldots, \left\{ Y_{i_N} \in A_{i_N} \right\} \subset \Omega,$$

are independent. Random variables $(Y_n)_{n\in\mathbb{N}}$ are *identically distributed* if

$$\mathbb{P}_{Y_1} = \mathbb{P}_{Y_2} = \mathbb{P}_{Y_3} = \dots .$$

A sequence of random variables $(Y_n)_{n\in\mathbb{N}}$ converges *in probability* to a random variable $Y\colon \Omega \to \mathcal{H}$ if we have

$$\mathbb{P}\left(d(Y, Y_n) > \varepsilon\right) \to 0, \quad \text{as } n \to \infty$$

for an arbitrary $\varepsilon > 0$.

7.2 Law of large numbers

Here we prove a nonlinear version of the law of large numbers and as an application we give an alternative proof of the Jensen inequality (Theorem 7.2.2). Another application is an algorithm for computing the Fréchet mean in the end of this section.

Let $(\Omega, \mathcal{F}, \mathbb{P})$ be a probability space and $(\mathcal{H}, d)$ be a Hadamard space. Given a sequence (Y_n) of random variables $Y_n\colon \Omega \to \mathcal{H}$, we define a new sequence (S_n) of random variables $S_n\colon \Omega \to \mathcal{H}$ as

$$S_1(\omega) := Y_1(\omega), \tag{7.2.4a}$$

and for $n \in \mathbb{N}$ inductively

$$S_{n+1}(\omega) := \frac{n}{n+1} S_n(\omega) + \frac{1}{n+1} Y_{n+1}(\omega), \tag{7.2.4b}$$

where $\omega \in \Omega$.

The following Theorem 7.2.1 states both the weak and strong laws of large numbers.

Theorem 7.2.1 (Law of large numbers). *Let (Y_n) be a sequence of independent identically distributed random variables $Y_n \in \mathcal{L}^2(\Omega, \mathcal{H})$ and let (S_n) be the sequence from (7.2.4). Then:*

(i) *The sequence (S_n) converges to $\mathbb{E}Y_1$ in $\mathcal{L}^2(\Omega, \mathcal{H})$ and in probability. (Weak law of large numbers)*

(ii) *If moreover $Y_n \in \mathcal{L}^\infty(\Omega, \mathcal{H})$ for each $n \in \mathbb{N}$, then*

$$S_n(\omega) \to \mathbb{E}Y_1, \quad \text{as } n \to \infty,$$

for $\mathbb{P}$-almost every $\omega \in \Omega$. (Strong law of large numbers)

Proof. Step 1: We claim that

$$\mathbb{E}d\left(Y_{n+1}, S_n\right)^2 \geq \mathbb{E}d\left(\mathbb{E}Y_{n+1}, S_n\right)^2 + \mathbb{E}d\left(\mathbb{E}Y_{n+1}, Y_{n+1}\right)^2 , \tag{7.2.5}$$

for every $n \in \mathbb{N}$. Indeed, observe that Y_{n+1} and S_n are independent and hence

$$\int_\Omega d\left(Y_{n+1}(\omega), S_n(\omega)\right)^2 \, d\mathbb{P}(\omega) = \int_\Omega \left[\int_\Omega d\left(Y_{n+1}(\omega), S_n\left(\omega'\right)\right)^2 \, d\mathbb{P}(\omega) \right] d\mathbb{P}\left(\omega'\right).$$

Fix $\omega' \in \Omega$ and apply (7.1.2) to obtain

$$\mathbb{E}d\left(Y_{n+1}, S_n\left(\omega'\right)\right)^2 \geq d\left(S_n\left(\omega'\right), \mathbb{E}Y_{n+1}\right)^2 + \mathbb{E}d\left(Y_{n+1}, \mathbb{E}Y_{n+1}\right)^2.$$

Integration with respect to ω' yields (7.2.5).

Step 2: Next we show by induction on $n \in \mathbb{N}$ that

$$\mathbb{E}d\left(\mathbb{E}Y_1, S_n\right)^2 \leq \frac{1}{n} \operatorname{var} Y_1 . \tag{7.2.6}$$

It obviously holds for $n = 1$ and we assume it holds for some fixed $n \in \mathbb{N}$. We have

$$\begin{aligned}
&\mathbb{E}d\left(\mathbb{E}Y_1, S_{n+1}\right)^2 \\
&\quad = \mathbb{E}d\left(\mathbb{E}Y_1, \frac{n}{n+1} S_n + \frac{1}{n+1} Y_{n+1}\right)^2 ,
\end{aligned}$$

by (1.2.2) we get

$$\leq \frac{n}{n+1} \mathbb{E}d\left(\mathbb{E}Y_1, S_n\right)^2 + \frac{1}{n+1} \mathbb{E}d\left(\mathbb{E}Y_1, Y_{n+1}\right)^2 - \frac{n}{(n+1)^2} \mathbb{E}d\left(Y_{n+1}, S_n\right)^2 ,$$

and applying (7.2.5) gives

$$\begin{aligned}
&\leq \frac{n}{n+1} \mathbb{E}d\left(\mathbb{E}Y_1, S_n\right)^2 + \frac{1}{n+1} \mathbb{E}d\left(\mathbb{E}Y_1, Y_{n+1}\right)^2 \\
&\quad - \frac{n}{(n+1)^2} \left[\mathbb{E}d\left(\mathbb{E}Y_{n+1}, S_n\right)^2 + \mathbb{E}d\left(\mathbb{E}Y_{n+1}, Y_{n+1}\right)^2\right] \\
&= \left(\frac{n}{n+1}\right)^2 \mathbb{E}d\left(\mathbb{E}Y_1, S_n\right)^2 + \frac{1}{(n+1)^2} \operatorname{var} Y_1 \\
&\leq \frac{1}{n+1} \operatorname{var} Y_1 .
\end{aligned}$$

This proves (7.2.6), which via Exercise 7.1 immediately yields (i).

Step 3: It remains to show (ii). We first claim that

$$S_{n^2}(\omega) \to \mathbb{E}Y_1, \quad \text{as } n \to \infty,$$

for $\mathbb{P}$-almost every $\omega \in \Omega$. Indeed, choose $\varepsilon > 0$ and apply (7.2.6) to arrive at

$$\sum_{n=1}^{\infty} \mathbb{P}\left(d\left(S_{n^2}, \mathbb{E}Y_1\right) > \varepsilon\right) \leq \sum_{n=1}^{\infty} \frac{1}{\varepsilon^2} \mathbb{E}d\left(S_{n^2}, \mathbb{E}Y_1\right)^2 \leq \sum_{n=1}^{\infty} \frac{1}{\varepsilon^2 n^2} \operatorname{var} Y_1 < \infty,$$

and the Borel–Cantelli lemma yields the claim. Now assume $Y_n \in \mathcal{L}^\infty(\Omega, \mathcal{H})$, that is, there exists $K > 0$ such that $d(z, Y_1(\omega)) \leq K$ for some $z \in \mathcal{H}$ and $\mathbb{P}$-almost every $\omega \in \Omega$. Then also $d(z, Y_n(\omega)) \leq K$ for every $n \in \mathbb{N}$ and $\mathbb{P}$-almost every $\omega \in \Omega$, since all the random variables are identically distributed, and by convexity also $d(z, S_n(\omega)) \leq K$ for every $n \in \mathbb{N}$ and $\mathbb{P}$-almost every $\omega \in \Omega$. Consequently,

$$d\left(S_n(\omega), S_{n+1}(\omega)\right) = \frac{1}{n+1} d\left(S_n(\omega), Y_{n+1}(\omega)\right) \leq \frac{2}{n+1} K,$$

for $\mathbb{P}$-almost every $\omega \in \Omega$. Therefore, for any $n, k \in \mathbb{N}$ with $n^2 \leq k < (n+1)^2$, we have

$$d\left(S_{n^2}(\omega), S_k(\omega)\right) \leq \left(\frac{1}{n^2+1} + \frac{1}{n^2+2} + \cdots + \frac{1}{k}\right) 2K \leq \frac{k - n^2}{n^2} 2K \leq \frac{4}{n} K,$$

which gives (ii). □

As a nice application of the law of large numbers, we reprove the Jensen inequality of Theorem 2.3.8.

Theorem 7.2.2 (Jensen inequality revisited). *Let $(\mathcal{H}, d)$ be a Hadamard space and $\mu \in \mathcal{P}^1(\mathcal{H})$. Then for a convex lsc function $f \colon \mathcal{H} \to (-\infty, \infty]$ we have*

$$f\left(b(\mu)\right) \leq \int_{\mathcal{H}} f(x)\, \mathrm{d}\mu(x).$$

Proof. Step 1: Let us first assume $\mu \in \mathcal{P}^2(\mathcal{H})$ and $f \in \mathcal{L}^2(\mathcal{H}, \mu)$. Choose a probability space $(\Omega, \mathcal{F}, \mathbb{P})$ and a sequence (Y_n) of independent random variables $Y_n \in \mathcal{L}^2(\Omega, \mathcal{H})$ such that $\mathbb{P}_{Y_n} = \mu$ for every $n \in \mathbb{N}$. Let S_n be defined by (7.2.4) and put

$$Z_n := f\left(Y_n\right), \quad \text{and} \quad T_n := \frac{1}{n} \sum_{i=1}^{n} Z_i,$$

for $n \in \mathbb{N}$. Then by the weak law of large numbers of Theorem 7.2.1 applied to (Y_n) and $(f \circ Y_n)$ we obtain

$$S_n \to \mathbb{E} Y_1 = b(\mu), \quad \text{and} \quad T_n \to \mathbb{E}\left(f \circ Y_1\right) = \int_{\mathcal{H}} f\, \mathrm{d}\mu, \tag{7.2.7}$$

in probability. We further prove by induction on $n \in \mathbb{N}$ that $f(S_n) \leq T_n$. Indeed, it is true for $n = 1$ and assume it is also true for some fixed $n \in \mathbb{N}$. By the convexity of f we have

$$f\left(S_{n+1}\right) \leq \frac{n}{n+1} f\left(S_n\right) + \frac{1}{n+1} f\left(Y_{n+1}\right) \leq \frac{n}{n+1} T_n + \frac{1}{n+1} Z_{n+1} = T_{n+1}.$$

There exist subsequences S_{n_k} and T_{n_k} such that the convergence in (7.2.7) is almost surely. Since f is lsc we get

$$f(b(\mu)) \leq \liminf_{k\to\infty} f\left(S_{n_k}\right) \leq \liminf_{k\to\infty} T_{n_k} = \int_{\mathcal{H}} f \, \mathrm{d}\mu.$$

In other words,

$$f\left(b\left(\mathbb{P}_{Y_1}\right)\right) \leq \mathbb{E}\left(f \circ Y_1\right). \tag{7.2.8}$$

Step 2: Let now $\mu \in \mathcal{P}^1(\mathcal{H})$. By Lemma 2.3.7 we may assume that $f^+ \in \mathcal{L}^1(\mathcal{H}, \mu)$.

Choose a probability space $(\Omega, \mathcal{F}, \mathbb{P})$ and a random variable $X\colon \Omega \to \mathcal{H}$ such that $\mathbb{P}_X = \mu$. Then $X \in \mathcal{L}^1(\Omega, \mathcal{H})$ as well as $(f \circ X)^+ \in \mathcal{L}^1(\Omega, \mathbb{P})$. Fix some $z \in \mathcal{H}$ and put

$$\Omega_n := \left\{\omega \in \Omega\colon d(z, X(\omega)) < n,\ \left|f(X(\omega))\right| < n\right\}.$$

Define $X_n \in \mathcal{L}^\infty(\Omega, \mathcal{H})$ with $f \circ X_n \in \mathcal{L}^\infty(\Omega)$ by

$$X_n(\omega) := \begin{cases} X(\omega) & \text{if } \omega \in \Omega_n \\ z & \text{if } \omega \in \Omega \setminus \Omega_n. \end{cases}$$

Then $X_n \to X$ in $\mathcal{L}^1(\Omega, \mathcal{H})$ and hence by Proposition 7.1.2 we obtain $\mathbb{E}X_n \to \mathbb{E}X$. Since f is lsc we arrive at

$$f(\mathbb{E}X) \leq \liminf_{n\to\infty} f\left(\mathbb{E}X_n\right).$$

Moreover,

$$\mathbb{E}(f \circ X) = \lim_{n\to\infty} \mathbb{E}\left[\chi_{\Omega_n}(f \circ X)\right] = \lim_{n\to\infty} \mathbb{E}\left[\chi_{\Omega_n}(f \circ X_n)\right] = \lim_{n\to\infty} \mathbb{E}(f \circ X_n)$$

and recalling (7.2.8) we get

$$\geq \liminf_{n\to\infty} f\left(\mathbb{E}X_n\right).$$

In summary,

$$f(b(\mu)) = f(\mathbb{E}X) \leq \mathbb{E}(f \circ X) = \int_{\mathcal{H}} f \, \mathrm{d}\mu,$$

which finishes the proof. □

As a second application of the law of large numbers, we show how to compute means.

Computing the Fréchet mean

Recall first the definition of a mean from Example 2.2.5. If $a_1, \ldots, a_N \in \mathcal{H}$ and $(w_1, \ldots, w_N) \in \Delta_{N-1}$, then

$$\operatorname*{arg\,min}_{x\in\mathcal{H}} \sum_{n=1}^{N} w_n d\left(x, a_n\right)^2,$$

exists, is unique and we call it the mean of $a_1, \ldots, a_N$. Except for a few trivial situations, it is not possible to compute a mean exactly and an approximation algorithm is desirable. We have already presented the splitting proximal point algorithm (Theorems 6.3.7 and 6.3.13) which is of course applies to computing means in locally compact Hadamard spaces. Here we introduce a different approximation algorithm based on the law of large numbers (Theorem 7.2.1), which however in some special cases coincides with the algorithm from Theorem 6.3.13, as will be discussed in detail in Remark 8.3.3. The approximation sequence is defined as follows.

Algorithm 7.2.3 (Computing a mean). At each step $i \in \mathbb{N}_0$, choose randomly $r_i \in \{1, \ldots, N\}$ according to the distribution $\overline{w} := (w_1, \ldots, w_N)$ and put

$$s_{i+1} := \frac{i}{i+1} s_i + \frac{1}{i+1} a_{r_i} . \tag{7.2.9}$$

The convergence of the sequence (s_i) to the mean of $a_1, \ldots, a_N$ is guaranteed by Theorem 7.2.1.

To explain the details, denote $[N] := \{1, \ldots, N\}$ and $\Omega := [N]^{\mathbb{N}_0}$. We define a σ-algebra on Ω as the power of the σ-algebra 2^N, and we define a probability measure on Ω as the power of the probability distribution $\overline{w}$, respectively. Then the random variables $Y_i \colon \Omega \to \{a_1, \ldots, a_N\}$ given by

$$Y_i(\omega) := a_{\omega_i} , \qquad \omega = (\omega_1, \omega_2, \ldots) \in \Omega ,$$

for each $i \in \mathbb{N}_0$, are independent and identically distributed, and satisfy the assumptions of Theorem 7.2.1. Define then the sequence of random variables (S_n) like in (7.2.4). Since $s_i = S_i(\omega)$ for almost every $\omega \in \Omega$ and every $i \in \mathbb{N}_0$, we obtain the desired convergence to

$$\mathbb{E}Y_1 = \operatorname*{arg\,min}_{x \in \mathcal{H}} \sum_{n=1}^{N} w_n d(x, a_n)^2 ,$$

where the equality follows immediately from the definition.

7.3 Conditional expectations

We continue by defining a conditional expectation and establishing its basic properties. It can be then used to develop the theory of stochastic processes and martingales in Hadamard spaces. Since this is already beyond the scope of our book, we at least give a number of references at the end of the present chapter. Our motivation is the Grohs–Wolfowitz theory and a consensus algorithm which follows from it. On the other hand a conditional expectation is interesting on its own and its definition makes use of metric projections onto convex subsets of a Hadamard space.

Scalar-valued random variables

Let us first recall the definition of a conditional expectation of a scalar-valued random variable. Let $(\Omega, \mathcal{F}, \mathbb{P})$ be a probability space. Assume $\mathcal{G} \subset \mathcal{F}$ is another σ-algebra. It is now important to distinguish whether a function is measurable with respect to $\mathcal{F}$ or $\mathcal{G}$. The restriction of $\mathbb{P}$ onto $\mathcal{G}$ is still denoted $\mathbb{P}$. Let $X\colon \Omega \to [-\infty, \infty]$ be a random variable from $\mathcal{L}^1(\Omega, \mathcal{F}, \mathbb{P})$, that is, an $\mathcal{F}$-measurable function with $\mathbb{E}|X| < \infty$. Then the function

$$\varphi(A) := \int_A X \, \mathrm{d}\mathbb{P}, \quad A \in \mathcal{G},$$

is a new measure on $\mathcal{G}$, which is absolutely continuous with respect to (the restriction of) the measure $\mathbb{P}$. The Radon–Nikodym theorem hence gives a $\mathcal{G}$-measurable function

$$\mathbb{E}_{\mathcal{G}} X := \frac{\mathrm{d}\varphi}{\mathrm{d}\mathbb{P}},$$

which is a unique $\mathcal{G}$-measurable function satisfying

$$\int_A \mathbb{E}_{\mathcal{G}} X \, \mathrm{d}\mathbb{P} = \int_A X \, \mathrm{d}\mathbb{P}, \tag{7.3.10}$$

for every $A \in \mathcal{G}$. This function $\mathbb{E}_{\mathcal{G}} X$ is called the *conditional expectation* of X with respect to $\mathcal{G}$. The conditional expectation

$$\mathbb{E}_{\mathcal{G}}\colon \mathcal{L}^1(\Omega, \mathcal{F}, \mathbb{P}) \to \mathcal{L}^1(\Omega, \mathcal{G}, \mathbb{P})$$

is a linear mapping with the following properties:

(i) If $X \in \mathcal{L}^1(\Omega, \mathcal{F}, \mathbb{P})$ and $Y \in \mathcal{L}^1(\Omega, \mathcal{G}, \mathbb{P})$, then $\mathbb{E}_{\mathcal{G}}(XY) = Y\mathbb{E}_{\mathcal{G}} X$. In particular, $\mathbb{E}_{\mathcal{G}}(Y) = Y$.

(ii) If $X \in \mathcal{L}^1(\Omega, \mathcal{F}, \mathbb{P})$ and $X \geq 0$, then $\mathbb{E}_{\mathcal{G}} X \geq 0$.

(iii) If $X \in \mathcal{L}^1(\Omega, \mathcal{F}, \mathbb{P})$ and $f\colon \mathbb{R} \to \mathbb{R}$ is convex with $f \circ X \in \mathcal{L}^1(\Omega, \mathcal{F}, \mathbb{P})$, then

$$f\left(\mathbb{E}_{\mathcal{G}} X\right) \leq \mathbb{E}_{\mathcal{G}}\left(f \circ X\right), \tag{7.3.11}$$

which is the Jensen inequality for conditional expectations.

(iv) If $X_n \to X$ in $\mathcal{L}^1(\Omega, \mathcal{F}, \mathbb{P})$, or in $\mathcal{L}^2(\Omega, \mathcal{F}, \mathbb{P})$, then $\mathbb{E}_{\mathcal{G}} X_n \to \mathbb{E}_{\mathcal{G}} X$ in $\mathcal{L}^1(\Omega, \mathcal{F}, \mathbb{P})$, or in $\mathcal{L}^2(\Omega, \mathcal{F}, \mathbb{P})$, respectively.

It is important that the conditional expectation of an L^2-random variable can be characterized in terms of metric projections.

Proposition 7.3.1. *Let $X \in \mathcal{L}^2(\Omega, \mathcal{F}, \mathbb{P})$ and $Z \in \mathcal{L}^2(\Omega, \mathcal{G}, \mathbb{P})$. Then*

$$\mathbb{E}\left|Y - \mathbb{E}_G X\right|^2 \leq \mathbb{E}|Y - Z|^2,$$

for every $Y \in \mathcal{L}^2(\Omega, \mathcal{F}, \mathbb{P})$. That is, the conditional expectation $\mathbb{E}_{\mathcal{G}} X$ is the metric projection of X onto the closed linear subspace $\mathcal{L}^2(\Omega, \mathcal{G}, \mathbb{P})$.

The statement in Proposition 7.3.1 can be used as the definition of the conditional expectation for L^2-functions and such a definition can be then extended for L^1-functions by a density argument. This motivates the following developments.

Conditional expectation in Hadamard spaces

We will now turn to Hadamard space valued random variables and define conditional expectations for them. Let $(\mathcal{H}, d)$ be and Hadamard space and $(\Omega, \mathcal{F}, \mathbb{P})$ be a probability space. Assume $\mathcal{G} \subset \mathcal{F}$ is another σ-algebra.

Given an $\mathcal{F}$-measurable mapping $Y\colon \Omega \to \mathcal{H}$, we define its *mean conditional variance* as

$$\operatorname{var}_{\mathcal{G}} Y := \inf \left\{ \mathbb{E} d(Y, Z)^2 \colon\ Z\colon \Omega \to \mathcal{H},\ \mathcal{G}\text{-measurable} \right\} . \tag{7.3.12}$$

Recalling (7.1.1) we see that the variance of Y is

$$\operatorname{var} Y := \inf_{z \in \mathcal{H}} \mathbb{E} d(Y, z)^2,$$

and is equal to the conditional variance with $\mathcal{G} = \{\emptyset, \Omega\}$. In general, we have $\operatorname{var}_{\mathcal{G}} Y \leq \operatorname{var} Y$.

In the present section, it is convenient to change the notation and denote the Lebesgue space $\mathcal{L}^p(\Omega, \mathcal{H})$ from Proposition 1.2.18 by $\mathcal{L}^p(\mathcal{F})$. Its subset consisting of $\mathcal{G}$-measurable mappings will be denoted $\mathcal{L}^p(\mathcal{G})$. Note that $\mathcal{L}^2(\mathcal{G})$ is a closed convex subset of the Hadamard space $\mathcal{L}^2(\mathcal{F})$.

Definition 7.3.2 (Conditional expectation). Let $Y \in \mathcal{L}^2(\mathcal{F})$. Its *conditional expectation* under $\mathcal{G}$ is defined as the metric projection of Y on the convex closed set $\mathcal{L}^2(\mathcal{G})$ and we denote it $\mathbb{E}_{\mathcal{G}}(Y)$.

Recall that conditional expectation is well defined due to Theorem 2.1.12, and (2.1.1) gives the following form of a variance inequality

$$d_2 (Z, Y)^2 \geq \operatorname{var}_{\mathcal{G}} Y + d_2 \left(Z, \mathbb{E}_{\mathcal{G}} Y\right)^2 ,$$

for each $Z \in \mathcal{L}^2(\mathcal{G})$. Observe that if $\mathcal{G} = \{\emptyset, \Omega\}$, we recover the usual expectation, that is, $\mathbb{E}_{\mathcal{G}} Y = \mathbb{E} Y$. Furthermore, by Proposition 7.3.1, we see that this definition gives the usual conditional expectation in the case of $\mathcal{H} = \mathbb{R}$. We can hence use the symbol $\mathbb{E}_{\mathcal{G}}$ for both.

Also, the following *conditional variance inequality* holds.

Proposition 7.3.3 (Conditional variance inequality). *Let $Y \in \mathcal{L}^2(\mathcal{F})$ and $Z \in \mathcal{L}^2(\mathcal{G})$. Then*

$$\mathbb{E}_{\mathcal{G}} d(Z, Y)^2 \geq \mathbb{E}_{\mathcal{G}} d \left(\mathbb{E}_{\mathcal{G}} Y, Y\right)^2 + d \left(\mathbb{E}_{\mathcal{G}} Y, Z\right)^2 , \tag{7.3.13}$$

holds $\mathbb{P}$-almost surely on Ω.

Proof. Step 1: Fix a set $A \in \mathcal{G}$ and put

$$d_A(Z_1, Z_2) := \left[\int_A d(Z_1(\omega), Z_2(\omega))^2 \, d\mathbb{P}(\omega) \right]^{\frac{1}{2}},$$

for every $\mathcal{G}$-measurable mappings $Z_1, Z_2 \colon \Omega \to \mathcal{H}$. Denote $\mathcal{L}^2(A, \mathcal{G})$ the quotient of $\mathcal{L}^2(\mathcal{G})$ with respect

$$Z_1 \sim Z_2 \quad \text{if } d_A(Z_1, Z_2) = 0.$$

Exercise 7.2 yields that $(\mathcal{L}^2(A, \mathcal{G}), d_A)$ is a Hadamard space. Since the function $d_A(\cdot, Y)^2$ is strongly convex on $\mathcal{L}^2(A, \mathcal{G})$, by Proposition 2.2.17 we have a unique minimizer $\mathbb{E}_{A,\mathcal{G}} Y \in \mathcal{L}^2(A, \mathcal{G})$ along with the variance inequality

$$d_A(\mathbb{E}_{A,\mathcal{G}} Y, Y)^2 + d_A(\mathbb{E}_{A,\mathcal{G}} Y, Z)^2 \le d_A(Z, Y)^2, \tag{7.3.14}$$

for every $Z \in \mathcal{L}^2(A, \mathcal{G})$. Likewise, for each $Z \in \mathcal{L}^2(\Omega \setminus A, \mathcal{G})$, we obtain

$$d_{\Omega\setminus A}(\mathbb{E}_{\Omega\setminus A,\mathcal{G}} Y, Y)^2 + d_{\Omega\setminus A}(\mathbb{E}_{\Omega\setminus A,\mathcal{G}} Y, Z)^2 \le d_{\Omega\setminus A}(Z, Y)^2. \tag{7.3.15}$$

Obviously,

$$\begin{aligned} d_2(\mathbb{E}_{\mathcal{G}} Y, Y)^2 &= \inf\left\{ d_\Omega(Z, Y)^2 \colon Z \in \mathcal{L}^2(\Omega, \mathcal{G}, Y) \right\} \\ &\ge \inf\left\{ d_A(Z, Y)^2 \colon Z \in \mathcal{L}^2(A, \mathcal{G}, Y) \right\} \\ &\quad + \inf\left\{ d_{\Omega\setminus A}(Z, Y)^2 \colon Z \in \mathcal{L}^2(\Omega \setminus A, \mathcal{G}, Y) \right\} \\ &= d_A(\mathbb{E}_{A,\mathcal{G}} Y, Y)^2 + d_{\Omega\setminus A}(\mathbb{E}_{\Omega\setminus A,\mathcal{G}} Y, Y)^2. \end{aligned} \tag{7.3.16}$$

Step 2: Now define a mapping $Y_A \colon \Omega \to \mathcal{H}$ by

$$Y_A := \begin{cases} \mathbb{E}_{A,\mathcal{G}} Y, & \text{on } A, \\ \mathbb{E}_{\Omega\setminus A,\mathcal{G}} Y, & \text{on } \Omega \setminus A. \end{cases}$$

Observe that $Y_A \in \mathcal{L}^2(\mathcal{G})$. The inequalities (7.3.14)–(7.3.16) imply that Y_A is a unique minimizer of the function $d_2(\cdot, Y)^2$ on $\mathcal{L}^2(\mathcal{G})$. Thus $Y_A = \mathbb{E}_{\mathcal{G}} Y$. In particular, $\mathbb{E}_{\mathcal{G}} Y = \mathbb{E}_{A,\mathcal{G}} Y$ almost surely on A. Inequality (7.3.14) then reads

$$\int_A d(Z, \mathbb{E}_{\mathcal{G}} Y)^2 \, d\mathbb{P} + \int_A d(\mathbb{E}_{\mathcal{G}} Y, Y)^2 \, d\mathbb{P} \le \int_A d(Z, Y)^2 \, d\mathbb{P},$$

for each $Z \in \mathcal{L}^2(\mathcal{G})$, or equivalently

$$d(Z, \mathbb{E}_{\mathcal{G}} Y)^2 + \mathbb{E}_{\mathcal{G}} d(\mathbb{E}_{\mathcal{G}} Y, Y)^2 \le \mathbb{E}_{\mathcal{G}} d(Z, Y)^2,$$

holds $\mathbb{P}$-almost surely on Ω, for each $Z \in \mathcal{L}^2(\mathcal{G})$, which finishes the proof. □

Theorem 7.3.4. *Let $X, Y \in \mathcal{L}^2(\mathcal{F})$. Then we have*

$$d(\mathbb{E}_{\mathcal{G}}X, \mathbb{E}_{\mathcal{G}}Y) \le \mathbb{E}_{\mathcal{G}}d(X,Y),$$

almost surely, and for every $p \in [1,\infty]$ also

$$d_p(\mathbb{E}_{\mathcal{G}}X, \mathbb{E}_{\mathcal{G}}Y) \le d_p(X,Y).$$

Proof. The conditional variance inequality (7.3.13) gives

$$d(\mathbb{E}_{\mathcal{G}}X, \mathbb{E}_{\mathcal{G}}Y)^2 \le \mathbb{E}_{\mathcal{G}}d(\mathbb{E}_{\mathcal{G}}Y, X)^2 - \mathbb{E}_{\mathcal{G}}d(\mathbb{E}_{\mathcal{G}}X, X)^2,$$

almost surely, as well as,

$$d(\mathbb{E}_{\mathcal{G}}X, \mathbb{E}_{\mathcal{G}}Y)^2 \le \mathbb{E}_{\mathcal{G}}d(\mathbb{E}_{\mathcal{G}}X, Y)^2 - \mathbb{E}_{\mathcal{G}}d(\mathbb{E}_{\mathcal{G}}Y, Y)^2.$$

Apply Corollary 1.2.5 to the points $X(\omega), Y(\omega), \mathbb{E}_{\mathcal{G}}X(\omega)$, and $\mathbb{E}_{\mathcal{G}}Y(\omega)$ to obtain

$$\begin{aligned} d(\mathbb{E}_{\mathcal{G}}X, Y)^2 + d(X, \mathbb{E}_{\mathcal{G}}Y)^2 \le{} & d(\mathbb{E}_{\mathcal{G}}X, X)^2 + d(\mathbb{E}_{\mathcal{G}}Y, Y)^2 \\ & + 2d(\mathbb{E}_{\mathcal{G}}X, \mathbb{E}_{\mathcal{G}}Y)\, d(X,Y), \end{aligned}$$

almost surely. Take conditional expectations and add up these inequalities to get

$$2d(\mathbb{E}_{\mathcal{G}}X, \mathbb{E}_{\mathcal{G}}Y)^2 \le 2d(\mathbb{E}_{\mathcal{G}}X, \mathbb{E}_{\mathcal{G}}Y)\, \mathbb{E}_{\mathcal{G}}d(X,Y),$$

almost surely, or,

$$d(\mathbb{E}_{\mathcal{G}}X, \mathbb{E}_{\mathcal{G}}Y) \le \mathbb{E}_{\mathcal{G}}d(X,Y),$$

which is the first claim. This in turn implies the second one,

$$\mathbb{E}d(\mathbb{E}_{\mathcal{G}}X, \mathbb{E}_{\mathcal{G}}Y)^p \le \mathbb{E}\left[\mathbb{E}_{\mathcal{G}}d(X,Y)\right]^p \le \mathbb{E}\left[\mathbb{E}_{\mathcal{G}}d(X,Y)^p\right] \le \mathbb{E}d(X,Y)^p,$$

where the middle inequality follows from (7.3.11). The proof is now complete. □

As a corollary we obtain that the definition of conditional expectation $\mathbb{E}_{\mathcal{G}}$ extends continuously from $\mathcal{L}^2(\mathcal{F})$ to $\mathcal{L}^1(\mathcal{F})$. Indeed, approximate a random variable $Y \in \mathcal{L}^1(\mathcal{F})$ by random variables $Y_n \in \mathcal{L}^\infty(\mathcal{F})$ defined as

$$Y_n(\omega) := \begin{cases} Y(\omega), & \text{if } d(Y(\omega), z) \le n, \\ z, & \text{otherwise}, \end{cases}$$

where $z \in \mathcal{H}$ is fixed. Then $Y_n \to Y$ in $\mathcal{L}^1(\mathcal{F})$. Theorem 7.3.4 implies that $(\mathbb{E}_{\mathcal{G}}Y_n)$ is a Cauchy sequence in $\mathcal{L}^1(\mathcal{G})$ and we denote its limit point $\mathbb{E}_{\mathcal{G}}Y$. One can conclude that a conditional expectation is well defined also on $\mathcal{L}^1(\mathcal{F})$. Moreover, both inequalities of Theorem 7.3.4 remain valid in $\mathcal{L}^1(\mathcal{F})$. Finally, given $p \in [1,\infty]$, the mapping $\mathbb{E}_{\mathcal{G}} : \mathcal{L}^p(\mathcal{F}) \to \mathcal{L}^p(\mathcal{G})$ is nonexpansive.

Filtered conditional expectations

Let $(\Omega, \mathcal{F}, \mathbb{P})$ be a probability space and $\mathcal{G}_1 \subset \mathcal{G}_2 \subset \mathcal{F}$ be other σ-algebras. If a random variable $X \colon \Omega \to [-\infty, \infty]$ is from $\mathcal{L}^1(\Omega, \mathcal{F}, \mathbb{P})$, that is, it is an $\mathcal{F}$-measurable function with $\mathbb{E}|X| < \infty$, then

$$\mathbb{E}_{\mathcal{G}_1} \mathbb{E}_{\mathcal{G}_2} X = \mathbb{E}_{\mathcal{G}_1} X . \tag{7.3.17}$$

On the other hand if $(\mathcal{H}, d)$ is a Hadamard space and $X \colon \Omega \to \mathcal{H}$ is from $\mathcal{L}^1(\mathcal{F})$, then (7.3.17) is no longer true in general; see Exercise 7.3. To overcome this noncommutativity issue, filtered conditional expectations are now introduced.

Let $(\Omega, \mathcal{F})$ a measurable space. A sequence $(\mathcal{F}_i)_{i \in \mathbb{N}_0}$ of σ-algebras on Ω is called a *filtration* if $\mathcal{F}_0 \subset \mathcal{F}_1 \subset \cdots \subset \mathcal{F}$. Then $(\Omega, \mathcal{F}, (\mathcal{F}_i))$ is a *filtered measurable space.* A filtered measurable space with a probability measure is called a *filtered probability space.*

Definition 7.3.5 (Filtered conditional expectation). Let $(\Omega, \mathcal{F}, (\mathcal{F}_i), \mathbb{P})$ be a filtered probability space and $Y \in \mathcal{L}^1(\mathcal{F}_l)$ for some $l \in \mathbb{N}_0$. Then we define the *filtered conditional expectation* by

$$\mathbb{E}\left(Y \| \mathcal{F}_k\right) := \mathbb{E}_{\mathcal{F}_k} \mathbb{E}_{\mathcal{F}_{k-1}} \ldots \mathbb{E}_{\mathcal{F}_{l-1}} Y , \quad k \in \mathbb{N}_0 ,$$

provided $k < l$. Otherwise we put $\mathbb{E}(Y \| \mathcal{F}_k) := Y$.

The filtered conditional expectation is a crucial building block for developing martingale theory in Hadamard space, but it is beyond the scope of the present book. We merely show the following version of the Jensen inequality, which is needed in the proof of Theorem 7.3.16.

Theorem 7.3.6 (Conditional Jensen inequality). *Let $(\mathcal{H}, d)$ be a Hadamard space and $(\Omega, \mathcal{F}, (\mathcal{F}_i)_{i \in \mathbb{N}_0}, \mathbb{P})$ be a filtered probability space. If $f \colon \mathcal{H} \to (-\infty, \infty]$ is a convex lsc function and $Y \in \mathcal{L}^1(\mathcal{F})$ with $\mathbb{E}(f \circ Y)^+ < \infty$, then*

$$f\left(\mathbb{E}_{\mathcal{F}_i} Y\right) \leq \mathbb{E}_{\mathcal{F}_i}(f \circ Y),$$

and if moreover $Y \in \mathcal{L}^1(\mathcal{F}_k)$, we have

$$f\left(\mathbb{E}\left(Y \| \mathcal{F}_i\right)\right) \leq \mathbb{E}_{\mathcal{F}_i}(f \circ Y) ,$$

for each $i, k \in \mathbb{N}_0$ with $i \leq k$.

Proof. Step 1: Assume first $Y \in \mathcal{L}^2(\mathcal{F})$ and $\mathbb{E}|f \circ Y|^2 < \infty$. The random variable $\widehat{Y} := (Y, f \circ Y)$ has values in the Hadamard space $\widehat{\mathcal{H}} := \mathcal{H} \times \mathbb{R}$ and it is an L^2-mapping satisfying

$$\mathbb{E}_{\mathcal{F}_i} \widehat{Y} = \left(\mathbb{E}_{\mathcal{F}_i} Y, \mathbb{E}_{\mathcal{F}_i}(f \circ Y)\right) .$$

Since $\widehat{Y}(\Omega) \subset \operatorname{epi} f$, we have $\mathbb{E}_{\mathcal{F}_i}\widehat{Y} \in \operatorname{epi} f$ and therefore

$$f\left(\mathbb{E}_{\mathcal{F}_i}Y\right) \le \mathbb{E}_{\mathcal{F}_i}(f \circ Y)\,.$$

Step 2: Assume now $Y \in \mathcal{L}^1(\mathcal{F})$ and $\mathbb{E}(f \circ Y)^+ < \infty$. Choose $z \in \mathcal{H}$ and put

$$\Omega_n := \left\{\omega \in \Omega\colon d\,(z, Y(\omega)) < n,\ \left|f\,(Y(\omega))\right| < n\right\}\,.$$

Define $Y_n \in \mathcal{L}^\infty(\Omega, \mathcal{H})$ with $f \circ Y_n \in \mathcal{L}^\infty(\Omega)$ by

$$Y_n(\omega) := \begin{cases} Y(\omega) & \text{if } \omega \in \Omega_n \\ z & \text{if } \omega \in \Omega \setminus \Omega_n. \end{cases}$$

Then $Y_n \to Y$ in $\mathcal{L}^1(\mathcal{F})$ and by Theorem 7.3.4 also $\mathbb{E}_{\mathcal{F}_i}Y_n \to \mathbb{E}_{\mathcal{F}_i}Y$ in $\mathcal{L}^1(\mathcal{F})$ as $n \to \infty$. The lower semicontinuity of f implies

$$\liminf_{n\to\infty} f\left(\mathbb{E}_{\mathcal{F}_i}Y_n\right) \ge f\left(\mathbb{E}_{\mathcal{F}_i}Y\right)\,.$$

Moreover,

$$\begin{aligned}
\mathbb{E}_{\mathcal{F}_i}(f \circ Y) &= \lim_{n\to\infty} \mathbb{E}_{\mathcal{F}_i}\left[\chi_{\Omega_n}(f \circ Y)\right] \\
&= \lim_{n\to\infty} \mathbb{E}_{\mathcal{F}_i}\left[\chi_{\Omega_n}(f \circ Y_n)\right] \\
&= \lim_{n\to\infty} \mathbb{E}_{\mathcal{F}_i}\left(f \circ Y_n\right) \\
&\ge \liminf_{n\to\infty} f\left(\mathbb{E}_{\mathcal{F}_i}Y_n\right)\,,
\end{aligned}$$

by Step 1.

Step 3: Finally, assume $Y \in \mathcal{L}^1(\mathcal{F}_k)$ and $\mathbb{E}(f \circ Y)^+ < \infty$. For $i \in \mathbb{N}$ with $i \le k$ define

$$Y_i := \mathbb{E}\left(Y \| \mathcal{F}_i\right)\,.$$

Using Step 2, we can iteratively deduce from $Y_i \in \mathcal{L}^1(\mathcal{F}_k)$ and $\mathbb{E}(f \circ Y_i)^+ < \infty$ that $Y_{i-1} \in \mathcal{L}^1(\mathcal{F}_k)$ and

$$f \circ Y_{i-1} \le \mathbb{E}_{\mathcal{F}_{i-1}}\left(f \circ Y_i\right)\,.$$

Iterating the last inequality finishes the proof. □

Grohs–Wolfowitz theory

We finish this chapter by presenting a remarkable result about a certain family of stochastic matrices and show its application to distributed computing. The following notation will be used.

Notation 7.3.7. Given $k \in \mathbb{N}$, we denote $[k] := \{1, \dots, k\}$. For a mapping $x\colon [k] \to \mathcal{H}$, the symbol $\operatorname{diam}(x)$ stands for the diameter of its range, that is, $\operatorname{diam}(x) := \max_{m,n\in[k]} d(x(m), x(n))$.

Recall some classical definitions.

Definition 7.3.8 (Markov kernels). For every $i, j \in \mathbb{N}_0$ with $i \le j$ let $p_{ij}\colon [N]\times[N] \to [0, 1]$ be a stochastic matrix. If the family $(p_{ij})_{ij}$ satisfies:
(i) $p_{ii}(m, n) = \delta_{mn}$, for every $i \in \mathbb{N}_0$ and $m, n = 1, \dots, N$,
(ii) $p_{ik} = p_{ij}p_{jk}$, for every $i, j, k \in \mathbb{N}_0$ with $i \le j \le k$,
then we say that (p_{ij}) form a family of *Markov kernels*.

Having a family of Markov kernels p_{ij}, where $i, j \in \mathbb{N}_0$ with $i \le j$, we define a Markov chain $(X_i)_{i\in\mathbb{N}_0}$ on the probability space $\Omega := [N]^{\mathbb{N}_0}$ equipped with the standard product σ-algebra $\mathcal{F}$. Given an initial distribution π on $[N]$ we equip the probability space $(\Omega, \mathcal{F})$ with the probability measure $\mathbb{P}$ which is uniquely determined by its restrictions onto the cylinders:

$$\mathbb{P}\left(\{k_0\} \times \cdots \times \{k_i\} \times [N] \times \dots\right) := \pi\left(k_0\right) p_{01}\left(k_0, k_1\right) \dots p_{i-1,i}\left(k_{i-1}, k_i\right) ,$$

where $k_0, \dots, k_i \in [N]$ and $i \in \mathbb{N}_0$. Then for each $i \in \mathbb{N}_0$, the random variable $X_i\colon \Omega \to [N]$ is given by

$$X_i(\omega) := \omega_i , \quad \omega := \left(\omega_0, \omega_1, \dots\right) \in \Omega .$$

This processes is adapted to the filtration

$$\mathcal{F}_i := \sigma\left\{\{k_0\} \times \cdots \times \{k_i\} \times [N] \times \dots :\ \left(k_0, \dots, k_i\right) \in [N]^{i+1}\right\} ,$$

where $i \in \mathbb{N}_0$, meaning that the random variable X_i is $\mathcal{F}_i$-measurable. If we define the expectation $\mathbb{E}$ is with respect to $\mathbb{P}$, then the *linear Markov property*, that is,

$$\mathbb{E}_{\mathcal{F}_i} f\left(X_j\right)(\omega) = \sum_{n=1}^{N} p_{ij}\left(X_i(\omega), n\right) f(n) , \quad \omega \in \Omega , \tag{7.3.18}$$

holds for an arbitrary function $f\colon [N] \to [0, \infty)$ and $i, j \in \mathbb{N}_0$ with $i \le j$.

Example 7.3.9 (Homogeneous Markov chain). Let $A \in \mathbb{R}^{N\times N}$ be a stochastic matrix and put $p_{i-1,i} := A$ for $i \in \mathbb{N}$. Then $p_{ij} = A^{j-i}$ are Markov kernels, where $i, j \in \mathbb{N}_0$ and $i \le j$, and the associated Markov process is called *homogeneous*.

Proposition 7.3.10 (Nonlinear Markov property). *Let p_{ij} be Markov kernels for $i, j \in \mathbb{N}_0$ with $i \le j$ and let $x\colon [N] \to \mathcal{H}$ be a mapping. Then we have*

$$\mathbb{E}_{\mathcal{F}_i}\left(x \circ X_j\right)(\omega) = \operatorname*{arg\,min}_{z\in\mathcal{H}} \sum_{n=1}^{N} p_{ij}\left(X_i(\omega), n\right) d\left(x(n), z\right)^2 ,$$

for every $i, j \in \mathbb{N}_0$ with $i \le j$. Here (X_i) is the Markov chain constructed above.

Proof. Step 1: We will first prove that

$$\int_\Omega \sum_{n=1}^N p_{ij}\left(X_i(\omega), n\right) f\left(x(n), Y(\omega)\right) \mathrm{d}\mathbb{P}(\omega) = \int_\Omega f\left(x \circ X_j(\omega), Y(\omega)\right) \mathrm{d}\mathbb{P}(\omega),$$

holds for each $Y\colon \Omega \to \mathcal{H}$ which is $\mathcal{F}_i$-measurable and each measurable function $f\colon \mathcal{H} \times \mathcal{H} \to [0, \infty)$. By a monotone class argument, we may assume $f(u, v) = g(u)h(v)$ for some measurable functions $g, h\colon \mathcal{H} \to [0, \infty)$. Then we obtain

$$\begin{aligned}
&\int_\Omega \sum_{n=1}^N p_{ij}\left(X_i(\omega), n\right) g\left(x(n)\right) h\left(Y(\omega)\right) \mathrm{d}\mathbb{P}(\omega) \\
&\qquad = \int_\Omega h\left(Y(\omega)\right) \sum_{n=1}^N p_{ij}\left(X_i(\omega), n\right) g\left(x(n)\right) \mathrm{d}\mathbb{P}(\omega),
\end{aligned}$$

and using the linear Markov property (7.3.18) we get

$$\begin{aligned}
&= \int_\Omega h\left(Y(\omega)\right) \mathbb{E}_{\mathcal{F}_i}\left[g\left(x \circ X_j\right)\right] \mathrm{d}\mathbb{P}(\omega) \\
&= \int_\Omega \mathbb{E}_{\mathcal{F}_i}\left[h\left(Y(\omega)\right) g\left(x \circ X_j\right)\right] \mathrm{d}\mathbb{P}(\omega) \\
&= \mathbb{E}\left[h(Y) g\left(x \circ X_j\right)\right],
\end{aligned}$$

which is already the desired equality.

Step 2: If we denote

$$Y(\omega) := \operatorname*{arg\,min}_{z \in \mathcal{H}} \sum_{n=1}^N p_{ij}\left(X_i(\omega), n\right) d\left(x(n), z\right)^2, \quad \omega \in \Omega,$$

we see that Y is $\mathcal{F}_i$-measurable, because X_i is such. It remains to show that $\mathbb{E}d(x \circ X_j, Y)^2 \le \mathbb{E}d(x \circ X_j, Z)^2$ for every $\mathcal{F}_i$-measurable random variable $Z\colon \Omega \to \mathcal{H}$. We obviously have

$$\sum_{n=1}^N p_{ij}\left(X_i, n\right) d\left(x(n), Y\right)^2 \le \sum_{n=1}^N p_{ij}\left(X_i, n\right) d\left(x(n), Z\right)^2,$$

and consequently

$$\begin{aligned}
\mathbb{E}d\left(x \circ X_j, Y\right)^2 &= \mathbb{E} \sum_{n=1}^N p_{ij}\left(X_i, n\right) d\left(x(n), Y\right)^2 \\
&\le \mathbb{E} \sum_{n=1}^N p_{ij}\left(X_i, n\right) d\left(x(n), Z\right)^2 \\
&= \mathbb{E}d\left(x \circ X_j, Z\right)^2.
\end{aligned}$$

The proof is complete. □

Remark 7.3.11. The expectation $\mathbb{E}$ of course depends on the initial probability distribution π, but Proposition 7.3.10 tells us that $\mathbb{E}_{\mathcal{F}_i}(x \circ X_j)(\omega)$ is *independent* of π.

A stochastic matrix $W \in \mathbb{R}^{N\times N}$ is *irreducible* if for every $m, n \in [N]$ there exists $i \in \mathbb{N}$ such that $W^i(m,n) > 0$. The matrix W is *aperiodic* if for every $m, n \in [N]$ we have $\gcd\{k \in \mathbb{N}\colon W^k(m,n) > 0\} = 1$. The following result is often referred to as the *Fundamental theorem of Markov chains.*

Theorem 7.3.12. *Let $W \in \mathbb{R}^{N\times N}$ be a stochastic matrix. It is irreducible and aperiodic if and only if the limit $Q := \lim_{i\to\infty} W^i$ exists and the matrix Q has all rows identical.*

A stochastic irreducible and aperiodic matrix W is called an *SIA matrix*. Note that $(Q(1,1),\dots,Q(1,N))$ is then a unique *stationary distribution* on $[N]$ for the homogeneous Markov chain from Example 7.3.9, that is, it is the only vector $q \in [0,1]^N$ satisfying $q = qW$.

Definition 7.3.13 (SIA family). A family $\{W_1,\dots,W_k\} \subset \mathbb{R}^{N\times N}$ is called an *SIA family* if every matrix

$$W_\eta := W_{\eta_1} W_{\eta_2} \dots W_{\eta_i}$$

is SIA, where $\eta \in [k]^i$ and $i \in \mathbb{N}$.

With regard to Theorem 7.3.12 it is natural to introduce the following pseudonorm

$$\|A\| := \max_{n\in[N]} \max_{k,m\in[N]} |A(k,n) - A(m,n)| ,$$

of a matrix $A \in \mathbb{R}^{N\times N}$. It quantifies how different the rows of a given matrix are. A sufficiently long word composed from an SIA family is a matrix with almost identical rows, as the following Wolfowitz theorem states.

Theorem 7.3.14. *Let $\{W_1,\dots,W_k\} \subset \mathbb{R}^{N\times N}$ be an SIA family. Then for every $\varepsilon > 0$ there exists $i_0 \in \mathbb{N}$ such that $\|W_\eta\| < \varepsilon$ whenever $\eta \in [k]^i$ for some $i \geq i_0$.*

Proof. See [206]. □

Example 7.3.15. Let $\{W_1,\dots,W_k\} \subset \mathbb{R}^{N\times N}$ be an SIA family and $\eta \in [k]^{\mathbb{N}}$. If we put $p_{i,i+1} := W_{\eta_i}$ for $i \in \mathbb{N}_0$, then we obtain Markov kernels and Theorem 7.3.14 tells us that the associated Markov chain converges to a stationary state.

We will now show that Theorem 7.3.14 can be extended into a Hadamard space $(\mathcal{H}, d)$ and applied as a consensus algorithm. Given a stochastic matrix $W \in \mathbb{R}^{N\times N}$ and a mapping $x\colon [N] \to \mathcal{H}$, we define

$$Wx(m) := \arg\min_{y\in\mathcal{H}} \sum_{n=1}^{N} W(m,n) d\left(y, x(n)\right)^2 ,$$

where $m = 1,\dots,N$.

Theorem 7.3.16. *Let $(\mathcal{H}, d)$ be a Hadamard space and $\{W_1, \dots, W_k\} \subset \mathbb{R}^{N\times N}$ be an SIA family. Then there exist $\kappa \in (0,1)$ and $i_0 \in \mathbb{N}$ such that we have*

$$\operatorname{diam}\left(W_\eta x\right) \le \kappa \operatorname{diam}(x)\,,$$

for every $\eta \in [k]^i$ with $i \ge i_0$ and every $x\colon [N] \to \mathcal{H}$.

Proof. We proceed in several steps. Step 1: We first show that

$$W_\eta x \circ X_0 = \mathbb{E}\left(x \circ X_i \| \mathcal{F}_0\right)\,, \tag{7.3.19}$$

for each $\eta \in [k]^i$ and $i \in \mathbb{N}$, with (X_i) being the Markov chain from Example 7.3.15. Indeed, by Proposition 7.3.10 we have

$$\begin{aligned}\mathbb{E}_{\mathcal{F}_{i-1}}\left(x \circ X_i\right)(\omega) &= \operatorname*{arg\,min}_{z\in\mathcal{H}} \sum_{n=1}^{N} W_{\eta_i}\left(X_{i-1}(\omega), j\right) d\left(z, x(n)\right)^2 \\ &= W_{\eta_i} x \circ X_{i-1}(\omega)\,,\end{aligned}$$

for each $\omega \in \Omega$, and repeating the argument yields

$$\begin{aligned}\mathbb{E}_{\mathcal{F}_{i-2}}\left(\mathbb{E}_{\mathcal{F}_{i-1}}\left(x \circ X_i\right)\right)(\omega) &= \mathbb{E}_{\mathcal{F}_{i-2}}\left(W_{\eta_i} x \circ X_{i-1}\right)(\omega) \\ &= W_{\eta_{i-1}} W_{\eta_i} x \circ X_{i-2}(\omega)\,,\end{aligned}$$

and easily also (7.3.19).

Step 2: Let $z \in \mathcal{H}$ and $i \in \mathbb{N}$. Using (7.3.19) gives

$$d\left(W_\eta x \circ X_0, z\right) = d\left(\mathbb{E}\left(x \circ X_i \| \mathcal{F}_0\right), z\right)$$

and if we apply Theorem 7.3.6, then

$$\le \mathbb{E}_{\mathcal{F}_0}\left[d\left(x \circ X_i, z\right)\right],$$

which by (7.3.18) gives

$$= \sum_{k=1}^{N} p_{0i}\left(X_0(\omega), k\right) d\left(z, x(k)\right)\,.$$

We hence arrive at

$$d\left(W_\eta x(m), z\right) \le \sum_{k=1}^{N} p_{0i}\left(m, k\right) d\left(z, x(k)\right) \tag{7.3.20}$$

for each $m \in [N]$. Put now $z := W_\eta x(n)$, with $n \in [N]$, and repeat the argument leading to (7.3.20) to obtain

$$d\left(W_\eta x(m), W_\eta x(n)\right) \le \sum_{k,l=1}^{N} p_{0i}\left(m, k\right) p_{0i}\left(n, l\right) d\left(x(k), x(l)\right)\,. \tag{7.3.21}$$

Step 3: The right-hand side (RHS) of (7.3.21) can be further estimated

$$
\begin{aligned}
\mathrm{RHS} &= \sum_{k,l=1}^{N} p_{0i}(1,k)\, p_{0i}(1,l)\, d\left(x(k), x(l)\right) \\
&\quad + \sum_{k,l=1}^{N} \left[p_{0i}(m,k) - p_{0i}(1,k)\right] p_{0i}(n,l)\, d\left(x(k), x(l)\right) \\
&\quad + \sum_{k,l=1}^{N} p_{0i}(1,k) \left[p_{0i}(n,l) - p_{01}(1,l)\right] d\left(x(k), x(l)\right) \\
&\leq \sum_{k,l=1}^{N} p_{0i}(1,k)\, p_{0i}(1,l)\, d\left(x(k), x(l)\right) + 2\left\|p_{0i}\right\| \operatorname{diam}(x) \\
&\leq \left(\sum_{k,l=\binom{1}{k}\neq l}^{N} p_{0i}(1,k)\, p_{0i}(1,l) + 2\left\|p_{0i}\right\| \right) \operatorname{diam}(x) \\
&= \left(1 - \sum_{k=1}^{N} p_{0i}(1,k)^2 + 2\left\|p_{0i}\right\| \right) \operatorname{diam}(x)\,.
\end{aligned} \tag{7.3.22}
$$

Now observe that

$$
\sum_{k=1}^{N} p_{0i}(1,k)^2 \geq \frac{1}{\sqrt{N}} \sum_{k=1}^{N} \left|p_{0i}(1,k)\right| = \frac{1}{\sqrt{N}}\,,
$$

and also, since $p_{0i} = W_{\eta_1} W_{\eta_2} \dots W_{\eta_i}$, there exists $i_0 \in \mathbb{N}$ by Theorem 7.3.14 such that

$$
\left\|p_{0i}\right\| \leq \frac{1}{4\sqrt{N}}\,,
$$

whenever $i \geq i_0$. Altogether, we arrive at

$$
d\left(W_\eta x(m), W_\eta x(n)\right) \leq \left(1 - \frac{1}{2\sqrt{N}}\right) \operatorname{diam}(x)\,,
$$

which finishes the proof. □

A consensus algorithm

We end this section by showing a possible application of Theorem 7.3.16. Assume we have a mapping $x\colon [N] \to \mathcal{H}$ and we would like to compute an average element of $x(1), \dots, x(N)$. It may happen that computing the geodesic between a certain pair of points $x(m)$ and $x(n)$ is difficult and we therefore cannot use Algorithm 7.2.3 to compute the mean. A possible solution to this issue is to find an SIA family $\{W_1, \dots, W_k\}$ such that each of its matrices reflects the possibility to compute the geodesic between $x(m)$ and $x(n)$ at a particular step. Then it suffices to put

$$
x_i(\cdot) := W_{\eta_i} x_{i-1}(\cdot)\,,
$$

for $i \in \mathbb{N}$ and Theorem 7.3.16 yields a point $\widehat{x} \in \mathcal{H}$ such that $x_i(n) \to \widehat{x}$ as $i \to \infty$ for every $n = 1, \ldots, N$. The limit point $\widehat{x}$ is a natural alternative to the mean if the latter is difficult to compute. Algorithms of this type are called *consensus algorithms*.

Exercises

Exercise 7.1. Let $Y_n \in \mathcal{L}^2(\Omega, \mathcal{H})$ be a sequence of random variables which converges to some $Y \in \mathcal{L}^2(\Omega, \mathcal{H})$ in the metric d_2. Show that then (Y_n) converges to Y in probability.

Exercise 7.2. Show that the space $(\mathcal{L}^2(A, \mathcal{G}), d_A)$ from the proof of Proposition 7.3.3 is Hadamard.
Hint. Mimic the proof of Proposition 1.2.18. □

Exercise 7.3. Show that (7.3.17) holds for functions, but not for mappings into a Hadamard space.
Hint. If need be, see [194, Example 3.2]. □

Bibliographical remarks

Probability theory in Hadamard spaces was developed by K.-T. Sturm. The material of this chapter comes from his papers [192, 194–196]. We follow the original text very closely. There are also other deep results, which were not mentioned in the present book, for instance, martingale theory [62, 194] and a probabilistic approach to harmonic maps [192, 193, 196].

Proposition 7.1.2 is a special case of [194, Theorem 2.3]. The law of large numbers appeared in [195, Theorem 4.7] and it was used to prove the Jensen inequality in [195, Theorem 6.2]. For computing the Fréchet mean via the law of large numbers, see [16, 143, 195].

Conditional expectation theory (Section 7.3) is from [194]. Its counterpart for scalar-valued random variables can be found in many books on probability, for instance [73]. The nonlinear Markov property (Proposition 7.3.10) is a special case of [194, Theorem 5.2], which was mentioned in a similar form in [75, Proposition 6] and used in [94, Proposition 2.6]. Theorem 7.3.16 is the main result of [94] and it extends Theorem 7.3.14 by J. Wolfowitz [206] into Hadamard spaces. For details on Markov chains and SIA matrices, see for instance [185].

The papers [15, 75, 80, 150] contain further results in probability in Hadamard spaces.

8 Tree space and its applications

Although the Hadamard space is an interesting object on its own and plays an important role in geometry and geometric group theory, we are pleased to present a Hadamard space which has real world applications. This space was constructed by L. Billera, S. Holmes and K. Vogtmann as a (moduli) space of metric trees with a fixed number of terminal vertices and is now called the *BHV tree space*, or just *tree space*. Since such trees typically model the evolutionary history of a group of species, tree space becomes a fitting framework for various computations in statistical phylogenetics.

We first present the construction of tree space and then, in Section 8.2, we will be concerned with an algorithm for computing a convex combination of a given pair of trees in tree space. Unlike the approximation algorithms in Chapter 6, this algorithm works in *finitely* many steps and returns an *exact* result. Finally, in Section 8.3 we give *approximation* algorithms for computing medians and means of trees and outline their applications in phylogenetics.

8.1 Construction of the BHV tree space

We now construct a space of trees, which will turn out to be a cubical CAT(0) complex. First we need to make precise what we mean by a tree. Given $n \in \mathbb{N}$, a *metric n-tree* is a tree[1] with $n+1$ terminal vertices called *leaves* that are labeled $0, 1, \dots, n$, such that the edges have positive lengths. Vertices other than leaves have no labels since we consider them just as 'branching points' and their degree is at least three. The edges which are adjacent to a leaf are called *leaf edges*, and the remaining edges are called *inner*. We see an example of a metric 6-tree with three inner edges e_1, e_2, and e_3 in Figure 8.1. Recall that all edges, both leaf and inner, have nonnegative lengths. Instead of a metric n-tree, we will write simply a *tree*. The number n will be fixed and clear from the context. Later, when we consider multiple trees, it will be important

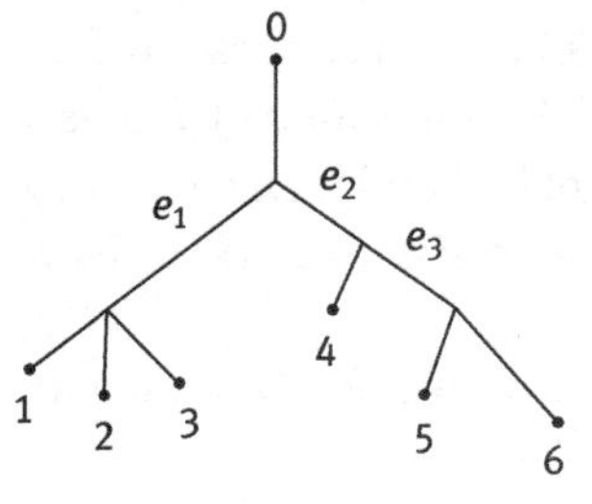

Fig. 8.1. An example of a 6-tree with three inner edges.

1 A tree is an undirected, connected, cycle-free graph.

that they all have the same number of leaves. The leaf 0 may in some applications play a distinguished role (for instance in phylogenetics it can represent a so-called *outgroup*), but in our considerations it has the same status as other leaves.

Each inner edge of a tree determines a unique partition of the set of leaves L into two disjoint and nonempty subsets $L_1 \cup L_2 = L$ called a *split*, which we denote $L_1|L_2$. A split is defined as the disjoint union of the two sets of leaves which arise when we removed the inner edge under consideration. For instance the inner edges e_1, e_2, e_3 of the tree in Figure 8.1 have splits $(0, 4, 5, 6|1, 2, 3)$, $(0, 1, 2, 3|4, 5, 6)$ and $(0, 1, 2, 3, 4|5, 6)$, respectively. On the other hand, having a set of leaves and splits subject to certain conditions, we can uniquely construct a tree [72, 184]. Namely, we require each pair of splits $L_1|L_2$ and $L'_1|L'_2$ be *compatible*, that is, one of the sets

$$L_1 \cap L'_1, \quad L_1 \cap L'_2, \quad L_2 \cap L'_1, \quad L_2 \cap L'_2$$

be empty. We say that a set of inner edges I is *compatible* if for each pair of edges $e, e' \in I$, the corresponding splits are compatible. These terms will be essential for computing convex combinations of a given pair of points in tree space in Section 8.2.

We will now proceed to construct a space of trees, that is, a space whose elements will be all metric n-trees, where $n \in \mathbb{N}$ is a fixed number representing the number of leaf vertices. The resulting space will be the tree space $\mathcal{T}_n$. First, it is useful to observe that one can treat separately leaf edges and inner edges. Indeed, since the lengths of the $n + 1$ leaf edges correspond to points in $[0, \infty)^{n+1}$, the space $\mathcal{T}_n$ is the product of $[0, \infty)^{n+1}$ and a factor representing the inner edges. We will now describe the latter. For simplicity, we may ignore the Euclidean factor of the space in the construction and denote the factor representing the inner edges also by $\mathcal{T}_n$.

Binary n-trees have the maximal possible number of inner edges, namely $n - 2$. Fix now a metric n-tree T with r inner edges of lengths $l_1, \ldots, l_r$, where $1 \le r \le n - 2$. Clearly $(l_1, \ldots, l_r)$ lies in the open orthant $(0, \infty)^r$, and conversely, each point of $(0, \infty)^r$ corresponds to an n-tree of the same combinatorial structure as T. Note that a tree T' is said to have the same combinatorial structure as T if it has the same number of inner edges as T and all its inner edges have the same splits as the inner edges of T. In other words, the trees T' and T differ only by inner edge lengths.

To each point from the boundary $\partial(0, \infty)^r$ we associate a metric n-tree obtained from T by shrinking some inner edges to zero length. Hence, each point from the closed orthant $[0, \infty)^r$ corresponds to a metric n-tree of the same combinatorial structure as T, where we allowed some inner edges to shrink to zero length.

An orthant containing trees that are not binary appears as a face of the orthants corresponding to (at least three) binary trees. In Figure 8.2, we see a copy of $[0, \infty)^2$ representing all 4-trees of a given combinatorial structure, namely, all 4-trees with two inner edges e_1 and e_2, such that the split of e_1 is $(1, 2|0, 3, 4)$ and the split of e_2 is $(1, 2, 3|0, 4)$. If the length of e_1 is zero, then the tree lies on the vertical boundary ray. If the length of e_2 is zero, then the tree lies on the horizontal boundary ray. In summary, each orthant $\mathcal{O} = [0, \infty)^r$, where $1 \le r \le n - 2$, corresponds to a compatible set

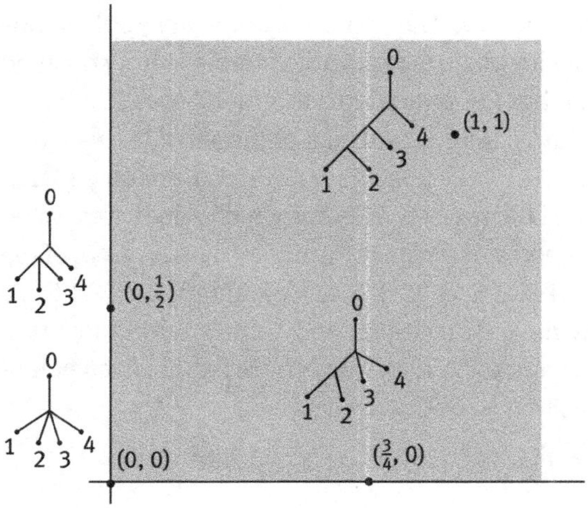

Fig. 8.2. 4-trees of a given combinatorial structure.

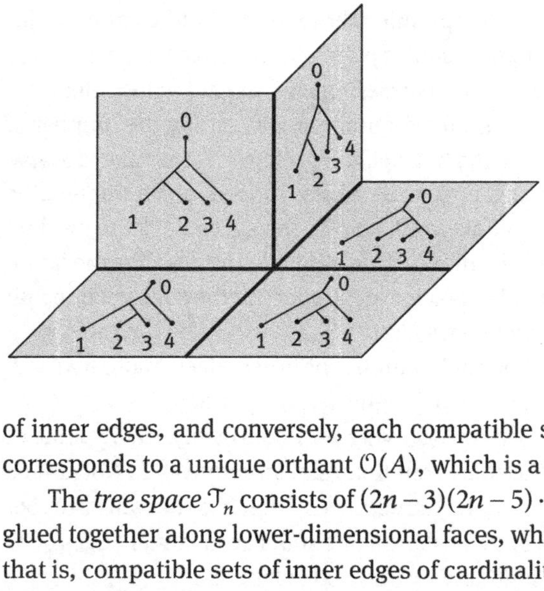

Fig. 8.3. Five out of 15 orthants of the tree space $\mathcal{T}_4$.

of inner edges, and conversely, each compatible set of inner edges $A := (e_1, \ldots, e_r)$ corresponds to a unique orthant $\mathcal{O}(A)$, which is a copy of $[0, \infty)^r$.

The *tree space* $\mathcal{T}_n$ consists of $(2n-3)(2n-5)\cdots 5\cdot 3$ copies of orthants $[0, \infty)^{n-2}$ glued together along lower-dimensional faces, which correspond to non-binary trees, that is, compatible sets of inner edges of cardinality $< n-2$. Note that the tree whose all inner edges have zero lengths can be view as the *origin of tree space*. We hence denote it 0.

We now equip the tree space $\mathcal{T}_n$ with a metric. Let $T, T' \in \mathcal{T}_n$. If both T and T' lie in the same orthant, we define their distance as the length of the Euclidean line segment $[T, T']$. If they lie in different orthants, we can connect them by a curve consisting of finitely many Euclidean line segments, and define the distance between T and T' as the infimum of the lengths of all such curves. Then $\mathcal{T}_n$ becomes a geodesic space. One can easily observe that each geodesic consists of finitely many Euclidean line segments each of which lies in one orthant. If we subdivide every orthant into the unit cubes having integral vertices, we obtain a cubical complex. The following important theorem states that tree space has nonpositive curvature.

Theorem 8.1.1. *The space $\mathcal{T}_n$ is a Hadamard space.*

Proof. We have already observed that $\mathcal{T}_n$ is a cubical complex. Indeed, it suffices to divide each orthant into cubes with integral vertices. By Theorem 1.2.14, it is a complete geodesic space. In order to apply Theorem 1.2.15, we need to show that the link of each vertex is a flag complex, which we leave as Exercise 8.1. Since $\mathcal{T}_n$ is obviously simply connected we obtain it is a Hadamard space. □

8.2 Owen–Provan algorithm

Next we are concerned with the issue of computing a convex combination of a given pair of trees $T, T' \in \mathcal{T}_n$ and computing their mutual distance. We will describe an ingenious algorithm which, given $t \in [0, 1]$, in a *finite* number of steps returns the tree $(1 - t)T + tT'$ and computes $d(T, T')$. From a practical point of view, it is important that this algorithm works in *polynomial* time. Our aim is to give a description detailed enough so that the reader understands the algorithm and is able to implement it on a computer. We however do not give proofs due to their technical nature and refer the reader to the original papers instead.

As in the construction of tree space in Section 8.1, we observe that the leaf edges of any tree on the geodesic between T and T' lie in the Euclidean factor of $\mathcal{T}_n$, and hence we can restrict our attention only to the inner edges. Denote $\mathcal{E}$ the set of inner edges of T and $\mathcal{E}'$ the set of inner edges of T'. If the trees T and T' have a common inner edge, that is, if there exist $e \in \mathcal{E}$ and $e' \in \mathcal{E}'$ which have the same splits, then this common edge will be present in any tree on the geodesic. This means that we can remove this edge and solve the original problem for the two subtrees which arise by removing this common edge. Hence we may assume without loss of generality that T and T' have no edge in common.

Let $\mathcal{A} := \{A_1, \dots, A_k\}$ and $\mathcal{B} := \{B_1, \dots, B_k\}$ be partitions of $\mathcal{E}$ and $\mathcal{E}'$, respectively. If

$$B_1 \cup \cdots \cup B_l \cup A_{l+1} \cup \cdots \cup A_k$$

is a compatible set for each $l \in \{1, \dots, k\}$, then there exist corresponding orthants $\mathcal{O}_l = \mathcal{O}_l(B_1 \cup \cdots \cup B_l \cup A_{l+1} \cup \cdots \cup A_k)$ in the tree space $\mathcal{T}_n$. The finite sequence of such orthants $\mathcal{P} := (\mathcal{O}_1, \dots, \mathcal{O}_k)$ is called a *path space* for the pair (T, T'). The shortest curve through a path space which connects T and T' is called a *path space geodesic*. The pair $(\mathcal{A}, \mathcal{B})$ is called its *support*. As the next result shows, it suffices to look for a geodesic among path space geodesics.

Proposition 8.2.1. *Let $T, T' \in \mathcal{T}_n$ be trees with no edge in common. Then the geodesic connecting T and T' is a path space geodesic for some path space between T and T'.*

We shall now proceed to identify path space geodesics. For a set of inner edges A, denote

$$\|A\| := \sqrt{\sum_{e \in A} |e|^2},$$

where $|e|$ stands for the length of e. It is sometimes important to specify in which tree a given edge is considered, we then use the symbol $|e|_T$ to denote the length of an edge e of a tree T.

Theorem 8.2.2. *Let $T, T' \in \mathcal{T}_n$ be trees with no edge in common. Then a curve $\gamma\colon [0,1] \to \mathcal{T}_n$ such that $\gamma(0) = T$ and $\gamma(1) = T'$ is a geodesic if and only if there exist partitions $\mathcal{A} = \{A_1, \dots, A_k\}$ and $\mathcal{B} = \{B_1, \dots, B_k\}$ of $\mathcal{E}$ and $\mathcal{E}'$, respectively, satisfying the following conditions:*

(i) *For any $l > m$, the sets A_l and B_m are compatible.*

(ii) *The sets satisfy*

$$\frac{\|A_1\|}{\|B_1\|} \le \frac{\|A_2\|}{\|B_2\|} \le \dots \le \frac{\|A_k\|}{\|B_k\|}.$$

(iii) *For each $l \in \{1, \dots, k\}$, there are no partition $C_1 \cup C_2$ of A_l, and partition $D_1 \cup D_2$ of B_l, such that C_2 is compatible with D_1 and*

$$\frac{\|C_1\|}{\|D_1\|} < \frac{\|C_2\|}{\|D_2\|},$$

and γ is a path space geodesic with support $(\mathcal{A}, \mathcal{B})$.

The algorithm for computing a geodesic is based on Theorem 8.2.2. We start with the support $(\mathcal{A}^0, \mathcal{B}^0)$, where $\mathcal{A} := \mathcal{E}$ and $\mathcal{B} := \mathcal{E}'$, and with the path space geodesic γ^0 which consists of the line segment connecting T and the origin of tree space 0 and the line segment connecting 0 and T'. Having a path space geodesic γ^i with support $(\mathcal{A}^i, \mathcal{B}^i)$ satisfying conditions (i) and (ii) of Theorem 8.2.2, we check whether the condition (iii) of Theorem 8.2.2 is also satisfied. If so, we have the geodesic, otherwise we construct a shorter path space geodesic γ^{i+1} with support $(\mathcal{A}^{i+1}, \mathcal{B}^{i+1})$ satisfying conditions (i) and (ii) of Theorem 8.2.2. By [157, Theorem 3.5] we know that this algorithm gives a geodesic in finitely many steps.

Let us now take a closer look at the iterative step $i \longrightarrow i+1$ and reformulate it as the extension problem for bipartite graphs. Given sets $A \subset \mathcal{E}$ and $B \subset \mathcal{E}'$, we define their *incompatibility graph* as a bipartite graph $G(A \cup B, E)$ with the vertex set $A \cup B$, whose edges E correspond to pairs $e \in A$ and $f \in B$ with incompatible splits. Clearly, two sets $C \subset A$ and $D \subset B$ are compatible if and only if $C \cup D$ is an *independent set* in $G(A \cup B, E)$. We define the weight of a vertex to be the squared length of the corresponding tree edge. The *extension problem* for $G(A \cup B, E)$ asks whether there exist a partition $C_1 \cup C_2$ of A and partition $D_1 \cup D_2$ of B such that $C_2 \cup D_1$ is an independent set in $G(A \cup B, E)$, and

$$\frac{\|C_1\|}{\|D_1\|} < \frac{\|C_2\|}{\|D_2\|}. \tag{8.2.1}$$

Hence a path space geodesic γ^i with support $(\mathcal{A}^i, \mathcal{B}^i)$ is a geodesic if and only if the extension problem has no solution for any pair (A_k, B_k) of $(\mathcal{A}^i, \mathcal{B}^i)$.

To reformulate the extension problem, we first note that scaling the edge lengths does not affect (8.2.1), hence we multiply each edge length so that $\|A\| = \|B\| = 1$. Then (8.2.1) is equivalent to

$$\|C_2\|^2 + \|D_1\|^2 > 1,$$

and we therefore look for an independent set in $G(A \cup B, E)$ having sufficiently large total weight. Since the sets C_2 and D_1 form an independent set if and only if their complements C_1 and D_2 form a *vertex cover* of the graph $G(A \cup B, E)$, the extension problem is equivalent to finding a vertex cover $C_1 \cup D_2 \subset A \cup B$ such that

$$\|C_1\|^2 + \|D_2\|^2 < 1.$$

For a general graph, this is a typical NP-hard problem (its decision version is NP-complete), but in case of bipartite graphs, it is classically solved via a flow network in polynomial time.

Indeed, we transform the bipartite graph $G(A \cup B, E)$, depicted in Figure 8.4, into a *flow network* as follows. First we impose an orientation to all edges from E so that they go from A to B, and assign infinite capacity to each of these edges. Then we add a new vertex called a *source*, and connect it with each vertex of A so that all these edges are oriented from the source to A. Set the capacity of an edge going from the source to a vertex $a \in A$ to be the weight of a. In a similar way, we add a new vertex called a *sink*, and connect all vertices of B with it. These edges are oriented from B to the sink and have capacities equal to the weights of the vertices of B. We hence obtained a flow network as in Figure 8.5. The aim now is to push as much of a flow from the source to the sink as possible. A maximal flow then gives a minimal cut by the *max-flow min-cut theorem*, and a minimal cut is exactly the desired vertex cover, as one can easily observe. There exist many algorithms which give a maximal flow in polynomial time. We recommend for instance the *push-relabel algorithm* due to A. Goldberg and R. Tarjan [93].

Recall that we consider two trees $T, T' \in \mathcal{T}_n$, which have no common edge, and our ultimate goal is to compute their distance and convex combination. So far we have found partitions $\mathcal{A} = \{A_1, \ldots, A_k\}$ and $\mathcal{B} = \{B_1, \ldots, B_k\}$ of the sets of edges $\mathcal{E}$ and $\mathcal{E}'$, respectively, satisfying the properties of Theorem 8.2.2. A path space geodesic $\gamma\colon [0,1] \to \mathcal{T}_n$ from T to T' with support $(\mathcal{A}, \mathcal{B})$ consists of $k+1$ Euclidean line

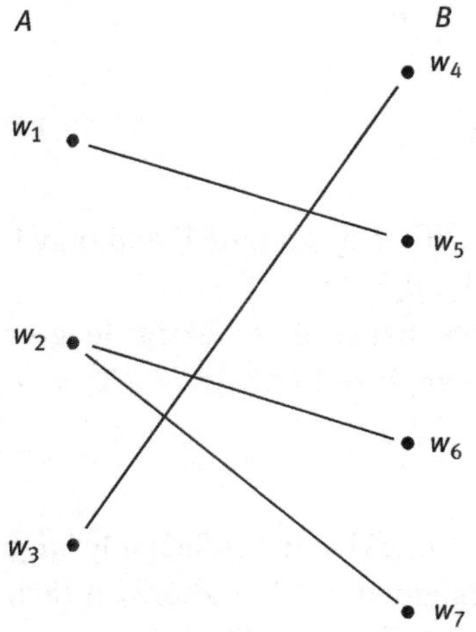

Fig. 8.4. A bipartite graph with vertex weights.

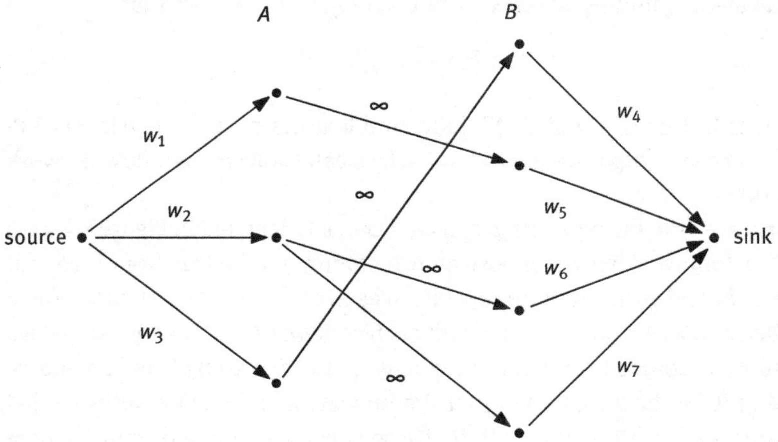

Fig. 8.5. The flow network with edge capacities.

segments

$$\begin{aligned}
&\left\{\gamma(t)\colon \tfrac{t}{1-t} \le \tfrac{\|A_1\|}{\|B_1\|}\right\}, && \text{if } i = 0,\\
&\left\{\gamma(t)\colon \tfrac{\|A_i\|}{\|B_i\|} \le \tfrac{t}{1-t} \le \tfrac{\|A_{i+1}\|}{\|B_{i+1}\|}\right\}, && \text{if } i = 1,\dots,k-1,\\
&\left\{\gamma(t)\colon \tfrac{\|A_k\|}{\|B_k\|}\right\}, && \text{if } i = k,
\end{aligned}$$

where $t \in [0,1]$, and where the points on each segment are associated with tree T_i having the edge set

$$B_1 \cup \cdots \cup B_i \cup A_{i+1} \cup \cdots \cup A_k,$$

and edge lengths

$$|e|_{T_i} := \begin{cases} \frac{(1-t)\|A_j\|-t\|B_j\|}{\|A_j\|}|e|_{T'}, & \text{if } e \in A_j, \\ \frac{t\|B_j\|-(1-t)\|A_j\|}{\|B_j\|}|e|_{T}, & \text{if } e \in B_j. \end{cases}$$

According to Theorem 8.2.2, this path space geodesic is actually a geodesic. Given now $t \in [0,1]$, it is straightforward to construct the tree $(1-t)T+tT'$. The distance between T and T' is given by

$$d(T,T') = \left[\sum_{i=1}^{k} (\|A_i\| + \|B_i\|)^2\right]^{\frac{1}{2}}.$$

If two trees have common edges, the decomposition into subtrees with no common edge can be done in linear time. Hence the whole Owen–Provan algorithm for finding a geodesic and computing distances in tree space works in polynomial time.

8.3 Medians and means of trees

In phylogenetics, the terminal vertices of a tree represent living species and the tree is to reflect the evolutionary history of these species. As a result of statistical computations (e.g. a Markov chain Monte Carlo simulation) or other methods, biologists obtain a large set of trees and it is desirable to represent this data set by a single tree. There are various reasons why the mean is a good choice. The task hence is, given a finite family of trees, to find their mean. This should serve as a brief motivation for the next developments.

Now we will show how to address the issue of computing a median and mean via the splitting proximal point algorithm as promised in Example 6.3.4. Let $(\mathcal{H}, d)$ be a locally compact Hadamard space. Assume $a_1, \ldots, a_N \in \mathcal{H}$ and $(w_1, \ldots, w_N) \in \Delta_{N-1}$. We minimize the function

$$f(x) := \sum_{n=1}^{N} w_n d(x, a_n)^p, \qquad x \in \mathcal{H}, \tag{8.3.2}$$

where $p = 1$ or $p = 2$, employing the splitting PPA with $f_n := w_n d(\cdot, a_n)^p$, for $n = 1, \ldots, N$. Given $\lambda > 0$, we denote the resolvent of f_n by J_λ^n, hence,

$$J_\lambda^n x := \operatorname*{arg\,min}_{y \in \mathcal{H}} \left[w_n d(y, a_n)^p + \frac{1}{2\lambda} d(y,x)^2\right], \qquad x \in \mathcal{H}, \tag{8.3.3}$$

and it is obvious that $J_\lambda^n x$ lies on the geodesic $[x, a_n]$, that is,

$$J_\lambda^n x = (1-t)x + t a_n,$$

for some $t \in [0,1]$ and our only task is to find such t. In other words we are to solve a *one-dimensional* problem at each step. We will now look at explicit forms of the approximation algorithms.

Algorithms for computing the mean

Let us first consider the cyclic order version from (6.3.19). Let (λ_k) be a sequence of positive reals satisfying (6.3.18). We start at some point $x_0 \in \mathcal{H}$ and for each $k \in \mathbb{N}_0$ we set

$$\begin{aligned} x_{kN+1} &:= J^1_{\lambda_k}\left(x_{kN}\right), \\ x_{kN+2} &:= J^2_{\lambda_k}\left(x_{kN+1}\right), \\ &\vdots \\ x_{kN+N} &:= J^N_{\lambda_k}\left(x_{kN+N-1}\right). \end{aligned}$$

It is easy to find the resolvents explicitly by (8.3.3). Indeed, fix $k \in \mathbb{N}_0$ and $1 \leq n \leq N$. Then x_{kN+n} lies on the geodesic $[x_{kN+n-1}, a_n]$ and by an elementary calculation we get

$$x_{kN+n} = \left(1 - t^n_k\right) x_{kN+n-1} + t^n_k a_n, \tag{8.3.4}$$

where

$$t^n_k = \frac{2\lambda_k w_n}{1 + 2\lambda_k w_n}. \tag{8.3.5}$$

The above algorithm then reads:

Algorithm 8.3.1 (Computing mean, cyclic order version). Given $x_0 \in \mathcal{H}$ and (λ_k) satisfying (6.3.18) we set

$$\begin{aligned} x_{kN+1} &:= \frac{1}{1 + 2\lambda_k w_1} x_{kN} + \frac{2\lambda_k w_1}{1 + 2\lambda_k w_1} a_1, \\ x_{kN+2} &:= \frac{1}{1 + 2\lambda_k w_2} x_{kN+1} + \frac{2\lambda_k w_2}{1 + 2\lambda_k w_2} a_2, \\ &\vdots \\ x_{kN+N} &:= \frac{1}{1 + 2\lambda_k w_N} x_{kN+N-1} + \frac{2\lambda_k w_N}{1 + 2\lambda_k w_N} a_N, \end{aligned}$$

for each $k \in \mathbb{N}_0$ and $n = 1, \ldots, N$.

The convergence of the sequence (x_j) produced by Algorithm 8.3.1 to the weighted mean of the points $a_1, \ldots, a_N$ follows by Corollary 6.3.15. Note that if the weights are uniform, that is, $w_n = \frac{1}{N}$ for each $n = 1, \ldots, N$, then the coefficients t^n_k are independent of n.

We will now turn to the randomized version from (6.3.25). By a similar process as above we obtain the following algorithm.

Algorithm 8.3.2 (Computing mean, random order version). Let $x_0 \in \mathcal{H}$ be a starting point and (λ_k) satisfy (6.3.24). At each step $k \in \mathbb{N}_0$, choose randomly $r_k \in \{1, \ldots, N\}$

according to the uniform distribution and put

$$x_{k+1} := \frac{1}{1 + 2\lambda_k w_{r_k}} x_k + \frac{2\lambda_k w_{r_k}}{1 + 2\lambda_k w_{r_k}} a_{r_k}.$$

The convergence of the sequence (x_k) produced by Algorithm 8.3.2 to the weighted mean of the points $a_1, \dots, a_N$ follows by Corollary 6.3.15.

Remark 8.3.3 (Comparing with Algorithm 7.2.3). We shall now compare Algorithm 8.3.2 with Algorithm 7.2.3. Let us first consider the *unweighted* case, that is, $w_n = \frac{1}{N}$ for every $n = 1, \dots, N$. At each iteration $k \in \mathbb{N}_0$, Algorithm 7.2.3 selects $r_k \in \{1, \dots, N\}$ according to the *uniform* distribution and generates a new point

$$s_{k+1} := \frac{k}{k+1} s_k + \frac{1}{k+1} a_{r_k}.$$

In Algorithm 8.3.2, at each step $k \in \mathbb{N}_0$ we again randomly choose a number $r_k \in \{1, \dots, N\}$ according to the *uniform* distribution and put

$$x_{k+1} := \frac{1}{1 + 2\lambda_k} x_k + \frac{2\lambda_k}{1 + 2\lambda_k} a_{r_k}.$$

Thus Algorithm 8.3.2 produces the same sequence as Algorithm 7.2.3 provided we set $\lambda_k := \frac{1}{2k}$ for each $k \in \mathbb{N}$. In other words Algorithm 7.2.3 is a special case of Algorithm 8.3.2.

On the other hand as far as *weighted* Fréchet means are concerned, there exists a difference between these two algorithms. Indeed, if $\overline{w} := (w_1, \dots, w_N)$ are the weights, then the algorithm (7.2.9) selects $r_k \in \{1, \dots, N\}$ according to the distribution $\overline{w}$ and generates a new point

$$s_{k+1} := \frac{k}{k+1} s_k + \frac{1}{k+1} a_{r_k},$$

that is, with the same coefficients as in the unweighted case. Algorithm 8.3.2 in contrast still selects $r_k \in \{1, \dots, N\}$ according to the *uniform* distribution, but the new point is given by

$$x_{k+1} = \frac{1}{1 + 2\lambda_k w_{r_k}} x_k + \frac{2\lambda_k w_{r_k}}{1 + 2\lambda_k w_{r_k}} a_{r_k},$$

that is, the coefficients now do depend on the weights. In summary, introducing weights effects either the coefficients (Algorithm 8.3.2), or the probability distribution which is used for selecting the points $a_1, \dots, a_N$ (Algorithm 7.2.3).

Algorithms for computing the median

We have to treat the cyclic and the random case separately. In the cyclic order version, we start at some point $x_0 \in \mathcal{H}$ and for each $k \in \mathbb{N}_0$ we set

$$\begin{aligned} x_{kN+1} &:= J^1_{\lambda_k}(x_{kN}), \\ x_{kN+2} &:= J^2_{\lambda_k}(x_{kN+1}), \\ &\vdots \\ x_{kN+N} &:= J^N_{\lambda_k}(x_{kN+N-1}), \end{aligned}$$

where $J^n_{\lambda_k}$ is the resolvent of the function $f_n := w_n d(\cdot, a_n)$, where $n = 1, \dots, N$, and (λ_k) is a sequence of positive reals satisfying (6.3.18). Given $k \in \mathbb{N}_0$ and $1 \le n \le N$, it is obvious that $J^n_{\lambda_k}(x_{kN+n-1})$ lies on the geodesic $[x_{kN+n-1}, a_n]$, that is,

$$x_{kN+n} = \left(1 - t^n_k\right) x_{kN+n-1} + t^n_k a_n,$$

for some $t^n_k \in [0, 1]$. These coefficients are again easy to determine.

Algorithm 8.3.4 (Computing median, cyclic order version). Given $x_0 \in \mathcal{H}$ and (λ_k) satisfying (6.3.18) we set

$$\begin{aligned} x_{kN+1} &:= \left(1 - t^1_k\right) x_{kN} + t^1_k a_1, \\ x_{kN+2} &:= \left(1 - t^2_k\right) x_{kN+1} + t^2_k a_2, \\ &\vdots \\ x_{kN+N} &:= \left(1 - t^N_k\right) x_{kN+N-1} + t^N_k a_N, \end{aligned}$$

with t^n_k defined by

$$t^n_k := \min\left\{1, \frac{\lambda_k w_n}{d\left(a_n, x_{kN+n-1}\right)}\right\},$$

for each $k \in \mathbb{N}_0$ and $n = 1, \dots, N$.

The convergence of the sequence (x_j) produced by Algorithm 8.3.4 to the median of the points $a_1, \dots, a_N$ with the weights $(w_1, \dots, w_N)$ follows by Corollary 6.3.15 above. Finally, the randomized version can be derived in a similar way.

Algorithm 8.3.5 (Computing median, random order version). Let $x_0 \in \mathcal{H}$ be a starting point and (λ_k) satisfy (6.3.24). At each step $k \in \mathbb{N}_0$, choose randomly $r_k \in \{1, \dots, N\}$ according to the uniform distribution and put

$$x_{k+1} := \left(1 - t_k\right) x_k + t_k a_{r_k}, \tag{8.3.6}$$

with t_k defined by

$$t_k := \min\left\{1, \frac{\lambda_k w_{r_k}}{d\left(a_{r_k}, x_k\right)}\right\},$$

for each $k \in \mathbb{N}_0$.

Table 8.1. Multiple sequence alignment of the 13 species.

Squirrel:	TAAGTAGAGGTGAAAAGCCTACCGAGCCTGGTGATAGCTGGTTGTCCAGACTA-GAATTTTAGTTCTACTTT
Baboon:	TGGGTAGAGGTGACAAGCCTACCGAGCCTGGTGATAGCTGGTTATCCAAGATA-GAATCTTAGTTCAGCCTT
Pika:	CAGATAGTGGTGAAAAGCCAAACGAGCCTAGTGATAGCTGGTTGTCCAGGTTA-GAATTTTAGTTCAACTTT
Guinea pig:	GAAGTAGAGGTGAAAAGCCAATCGAGCTTGGTGATAGCTGGTTATCCAAATAA-GAATATCAGTTCAGCTTT
Marmoset:	TGGGTAGTGGTGACAAACCTACCGAGCCTGGTGATAGCTGGTTATCCAAGACA-GAATCTTAGTTCAACTTT
Orangutan:	TGGGTAGAGGCGACAAACCTACCGAGCCTGGTGATAGCTGGTTGTCCAAGACA-GAATCTTAGTTCAACTTT
Mouse:	TAGGTAGAGGTGAAAAGCCTAACGAGCTTGGTGATAGCTGGTTACCCAAAAAATGAATTTAAGTTCAATTTT
Human:	TAGGTAGAGGCGACAAACCTACCGAGCCTGGTGATAGCTGGTTGTCCAAGATA-GAATCTTAGTTCAACTTT
Rat:	TAGGTAGAGGTGAAAAGCCTATCGAGCTTGGTGATAGCTGGTTGCCCAAAAAA-GAATTTCAGTTCAAACTT
Rabbit:	TAAATAGAGGTGAAAAGCCAACCGAGCCTGGTGATAGCTGGTTGTCCAGAATA-GAATTTTAGTTCAACTTT
Chimpanzee:	TAGGTAGAGGCGACAAACCTACCGAGCCTGGTGATAGCTGGTTGTCCAAGATA-GAATCTTAGTTCAACTTT
Kangaroo rat:	AAGGTAGAGGTGAAAAGCCTAACGAGCCTGATGATAGCTGGTTGTCCAAGAAA-GAATCTTAGTTCACCTTA
Gorilla:	TAGGTAGAGGCGACAAACCTACCGAGCCTGGTGATAGCTGGTTGTCCAAGACA-GAATCTTAGTTCAACTTT

The convergence of the sequence (x_k) produced by Algorithm 8.3.5 to the median of the points $a_1, \dots, a_N$ follows by Corollary 6.3.15.

A real data computation

We can apply the preceding algorithms in tree space $\mathcal{T}_n$ to compute the median and mean of a family of trees $T_1, \dots, T_N \in \mathcal{T}_n$. In this case the Owen–Provan algorithm is needed to compute convex combinations of a given pair of point. For instance, in Algorithm 8.3.1 we have to compute the convex combination in (8.3.4) and as a matter of fact, this is done at each iteration of this algorithm. It is therefore important that the Owen–Provan algorithm works in polynomial time, which makes Algorithm 8.3.1 fast and usable in practice. Likewise for the remaining algorithms for computing medians and means.

To demonstrate a real data computation, we present the following example from phylogenetics. The labels $\{0, \dots, n\}$ of a (phylogenetic) tree now represent a group of living species. In our example 13 species are considered and hence $n = 12$. In Table 8.1 we list the species along with a piece of their DNA sequences[2] called a *multiple sequence alignment*. There are several methods for constructing trees out of a multiple sequence alignment and we refer the interested reader to [34] for a description of one of such statistical models relying upon a *Markov chain Monte Carlo* simulation.[3] As a result we obtain sample trees from a posterior distribution in the tree space $\mathcal{T}_{12}$. The mean of the generated trees is an approximation of the posterior expectation and the higher the number of sample trees, the more accurate this approximation. In our computation we generated 100,000 such trees and depict four of them in Figure 8.6. Looking at, for instance, the position of the *guinea pig*, one can easily observe that the trees have different topologies and we therefore really need to use multiple orthants

2 This particular data comes from the MT-RNR2 gene.

3 For a detailed background, see [159].

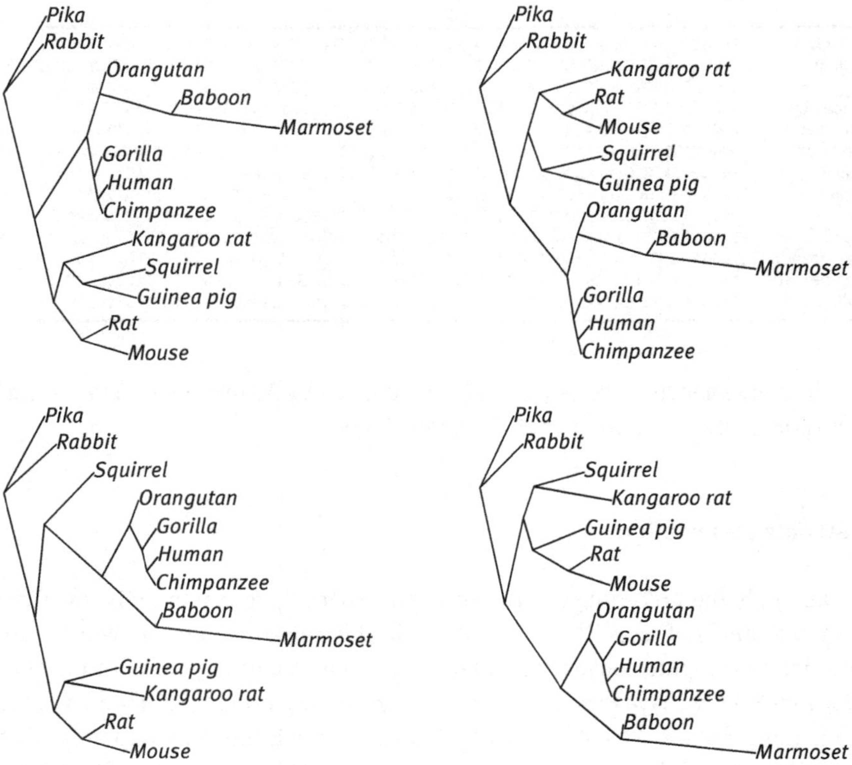

Fig. 8.6. Four out of the 100,000 trees in our example.

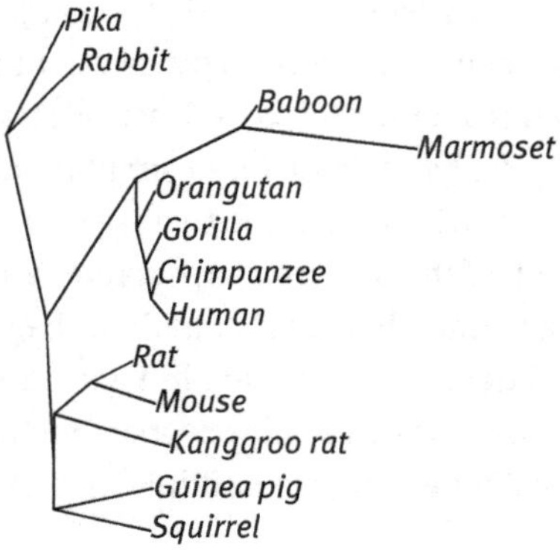

Fig. 8.7. The mean of the 100,000 trees in our example.

of tree space and hence the full strength of Hadamard space optimization. The mean of these 100,000 trees was computed using Algorithm 8.3.2 with step sizes $\lambda_k := \frac{1}{k+1}$, where $k \in \mathbb{N}_0$. The result of 10^8 iterations of this algorithm is depicted in Figure 8.7. This tree is a good point estimate of the posterior distribution.

Exercises

Exercise 8.1. Complete the missing bit in the proof of Theorem 8.1.1 by showing that the link of each vertex is a flag complex.
Hint. See [42, Lemma 4.1]. □

Bibliographical remarks

Tree space was introduced in the seminal paper [42], where is was also shown that it is a Hadamard space. For a long time there was no efficient algorithm for computing convex combinations of a given pair of points, which made the BHV tree space less attractive for applications. This situation has changed after the discovery of M. Owen and S. Provan [157]. The Owen–Provan algorithm builds upon prior studies of geodesics in tree space, which were worked out in [42, 158]. Proposition 8.2.1 was proved in [42, Proposition 4.1]; see also [157, Theorem 2.2]. Theorem 8.2.2 contains results from [158] and [157, Theorem 2.5].

The algorithms for computing medians and means appeared in [16]. Their implementations are available as free software [20, 35]. A statistical model for phylogenetic inference which relies upon these algorithms was developed in [34].

As a reference for graph theory needed in this chapter, we used [161].

We also recommend the excellent book [159] by L. Pachter and B. Sturmfels for the mathematical background of contemporary phylogenetics.

Since a tree-like structure is frequent in nature, tree space can be used as a framework for various biological models and computing medians and means is often of importance. In this way tree space has been used to model brain arteries [202] and lung airways [83, 84].

The space $P(n, \mathbb{R})$ of symmetric positive definite matrices $n \times n$ with real entries is a Hadamard manifold; see Example 1.2.12. This manifold plays a key role in *diffusion tensor imaging* as explained in [165] and computing Fréchet means of a finite family of symmetric positive definite matrices is one of the crucial operations [165, Section 3.7]. The algorithms of Section 8.3 can therefore find applications also in this area.

The real data computations and visualizations in this chapter are courtesy of Philipp Benner. I am most grateful to Megan Owen for her very valuable comments on this chapter.

Martin Bridson kindly pointed out to me that the BHV tree space is also a very interesting object in geometric group theory: it is essentially the cone over the link of the *rose* in the simplicial spine of *Outer space* introduced in [69] by M. Culler and K. Vogtmann. The BHV construction shows that there is a natural CAT(0) metric on this cone, but in contrast there does *not* exist a CAT(0) metric on the whole of the spine of Outer space. This result is due to M. Bridson [50]. For a gentle introduction to Culler–Vogtmann's Outer space, we recommend [201].

References

[1] P. Abramenko and K. S. Brown, *Buildings*, Graduate Texts in Mathematics 248, Springer, New York, 2008, Theory and applications.

[2] A. D. Aleksandrov, *A theorem on triangles in a metric space and some of its applications*, Trudy Mat. Inst. Steklov., v 38, Trudy Mat. Inst. Steklov., v 38, Izdat. Akad. Nauk SSSR, Moscow, 1951, pp. 5–23.

[3] S. Alexander, V. Kapovich and A. Petrunin, *Alexandrov geometry*, Book in preparation.

[4] S. B. Alexander and R. L. Bishop, The Hadamard-Cartan theorem in locally convex metric spaces, *Enseign. Math. (2)* 36 (1990), 309–320.

[5] S. B. Alexander and R. L. Bishop, Warped products of Hadamard spaces, *Manuscripta Math.* 96 (1998), 487–505.

[6] A. D. Alexandrow, *Die innere Geometrie der konvexen Flächen*, Akademie-Verlag, Berlin, 1955.

[7] L. Ambrosio and N. Gigli, *A user's guide to optimal transport*, Modelling and optimisation of flows on networks, Lecture Notes in Math. 2062, Springer, Heidelberg, 2013, pp. 1–155.

[8] L. Ambrosio, N. Gigli and G. Savaré, *Gradient flows in metric spaces and in the space of probability measures*, second ed, Lectures in Mathematics ETH Zürich, Birkhäuser Verlag, Basel, 2008.

[9] L. Ambrosio and G. Savaré, *Gradient flows of probability measures*, Handbook of differential equations: evolutionary equations. Vol. III, Handb. Differ. Equ., Elsevier/North-Holland, Amsterdam, 2007, pp. 1–136.

[10] K. Aoyama, Y. Kimura, W. Takahashi and M. Toyoda, Approximation of common fixed points of a countable family of nonexpansive mappings in a Banach space, *Nonlinear Anal.* 67 (2007), 2350–2360.

[11] D. Ariza-Ruiz, L. Leuştean and G. López-Acedo, Firmly nonexpansive mappings in classes of geodesic spaces, *Trans. Amer. Math. Soc.* 366 (2014), 4299–4322.

[12] H. Attouch, *Variational convergence for functions and operators*, Applicable Mathematics Series, Pitman (Advanced Publishing Program), Boston, MA, 1984.

[13] H. Attouch, Familles d'opérateurs maximaux monotones et mesurabilité, *Ann. Mat. Pura Appl. (4)* 120 (1979), 35–111.

[14] A. Auslender, *Méthodes numériques pour la résolution des problèmes d'optimisation avec constraintes*, Ph.D. thesis, Faculté des Sciences, Grenoble, 1969.

[15] T. Austin, A $\mathrm{CAT}(0)$-valued pointwise ergodic theorem, *J. Topol. Anal.* 3 (2011), 145–152.

[16] M. Bačák, Computing medians and means in Hadamard spaces, *to appear in SIAM J. Optim. arXiv:1210.2145*.

[17] M. Bačák, Convergence of semigroups under nonpositive curvature, *to appear in Trans. Amer. Math. Soc., arXiv:1211.0414*.

[18] M. Bačák, A new proof of the Lie-Trotter-Kato formula in Hadamard spaces, *appeared in Commun. Contemp. Math., http://www.worldscientific.com/doi/abs/10.1142/S0219199713500442*.

[19] M. Bačák, The proximal point algorithm in metric spaces, *Israel J. Math.* 194 (2013), 689–701.

[20] M. Bačák, *TrAP – Tree Averaging Program: https://github.com/bacak/TrAP*, 2013.

[21] M. Bačák and S. Reich, The asymptotic behavior of a class of nonlinear semigroups in Hadamard spaces, *accepted in J. Fixed Point Theory Appl., arXiv:1405.6637*.

[22] M. Bačák, I. Searston and B. Sims, Alternating projections in $\mathrm{CAT}(0)$ spaces, *J. Math. Anal. Appl.* 385 (2012), 599–607.

[23] J. B. Baillon, Un exemple concernant le comportement asymptotique de la solution du problème $du/dt + \partial\varphi(u) \ni 0$, *J. Funct. Anal.* 28 (1978), 369–376.

[24] J. B. Baillon, R. E. Bruck and S. Reich, On the asymptotic behavior of nonexpansive mappings and semigroups in Banach spaces, *Houston J. Math.* 4 (1978), 1–9.

[25] W. Ballmann, *Lectures on spaces of nonpositive curvature*, DMV Seminar 25, Birkhäuser Verlag, Basel, 1995, With an appendix by Misha Brin.
[26] W. Ballmann, M. Gromov and V. Schroeder, *Manifolds of nonpositive curvature*, Progress in Mathematics 61, Birkhäuser Boston Inc., Boston, MA, 1985.
[27] S. Banert, Backward–backward splitting in Hadamard spaces, *J. Math. Anal. Appl.* 414 (2014), 656–665.
[28] H. H. Bauschke and J. M. Borwein, On the convergence of von Neumann's alternating projection algorithm for two sets, *Set-Valued Anal.* 1 (1993), 185–212.
[29] H. H. Bauschke, J. V. Burke, F. R. Deutsch, H. S. Hundal and J. D. Vanderwerff, A new proximal point iteration that converges weakly but not in norm, *Proc. Amer. Math. Soc.* 133 (2005), 1829–1835 (electronic).
[30] H. H. Bauschke, Fenchel duality, Fitzpatrick functions and the extension of firmly nonexpansive mappings, *Proc. Amer. Math. Soc.* 135 (2007), 135–139 (electronic).
[31] H. H. Bauschke and P. L. Combettes, *Convex analysis and monotone operator theory in Hilbert spaces*, CMS Books in Mathematics/Ouvrages de Mathématiques de la SMC, Springer, New York, 2011, With a foreword by Hédy Attouch.
[32] H. H. Bauschke, E. Matoušková and S. Reich, Projection and proximal point methods: convergence results and counterexamples, *Nonlinear Anal.* 56 (2004), 715–738.
[33] H. H. Bauschke and X. Wang, *Firmly nonexpansive and Kirszbraun-Valentine extensions: a constructive approach via monotone operator theory*, Nonlinear analysis and optimization I. Nonlinear analysis, Contemp. Math. 513, Amer. Math. Soc., Providence, RI, 2010, pp. 55–64.
[34] P. Benner, M. Bačák and P. Y. Bourguignon, Point estimates in phylogenetic reconstructions, *To appear in Bioinformatics*.
[35] P. Benner, *TFBayes: https://github.com/pbenner/tfbayes*, 2013.
[36] G. C. Bento, O. P. Ferreira and P. R. Oliveira, Local convergence of the proximal point method for a special class of nonconvex functions on Hadamard manifolds, *Nonlinear Anal.* 73 (2010), 564–572.
[37] Y. Benyamini and J. Lindenstrauss, *Geometric nonlinear functional analysis. Vol. 1*, American Mathematical Society Colloquium Publications 48, American Mathematical Society, Providence, RI, 2000.
[38] V. N. Berestovskij and I. G. Nikolaev, *Multidimensional generalized Riemannian spaces*, Geometry, IV, Encyclopaedia Math. Sci. 70, Springer, Berlin, 1993, pp. 165–243, 245–250.
[39] I. D. Berg and I. G. Nikolaev, Quasilinearization and curvature of Aleksandrov spaces, *Geom. Dedicata* 133 (2008), 195–218.
[40] D. P. Bertsekas, Incremental proximal methods for large scale convex optimization, *Math. Program.* 129 (2011), 163–195.
[41] D. P. Bertsekas and J. N. Tsitsiklis, *Neuro-Dynamic Programming*, Athena Scientific, 1996.
[42] L. J. Billera, S. P. Holmes and K. Vogtmann, Geometry of the space of phylogenetic trees, *Adv. in Appl. Math.* 27 (2001), 733–767.
[43] R. L. Bishop and B. O'Neill, Manifolds of negative curvature, *Trans. Amer. Math. Soc.* 145 (1969), 1–49.
[44] V. S. Borkar, *Probability theory*, Universitext, Springer-Verlag, New York, 1995, An advanced course.
[45] J. Borwein, S. Reich and I. Shafrir, Krasnoselski-Mann iterations in normed spaces, *Canad. Math. Bull.* 35 (1992), 21–28.
[46] H. Brézis, *Opérateurs maximaux monotones et semi-groupes de contractions dans les espaces de Hilbert*, North-Holland Publishing Co., Amsterdam, 1973.
[47] H. Brézis and P. L. Lions, Produits infinis de résolvantes, *Israel J. Math.* 29 (1978), 329–345.

[48] H. Brezis and A. Pazy, Semigroups of nonlinear contractions on convex sets, *J. Functional Analysis* 6 (1970), 237–281.

[49] H. Brézis and A. Pazy, Convergence and approximation of semigroups of nonlinear operators in Banach spaces, *J. Functional Analysis* 9 (1972), 63–74.

[50] M. R. Bridson, *Geodesics and curvature in metric simplicial complexes*, Group theory from a geometrical viewpoint (Trieste, 1990), World Sci. Publ., River Edge, NJ, 1991, pp. 373–463.

[51] M. R. Bridson and A. Haefliger, *Metric spaces of non-positive curvature*, Grundlehren der Mathematischen Wissenschaften [Fundamental Principles of Mathematical Sciences] 319, Springer-Verlag, Berlin, 1999.

[52] R. E. Bruck and S. Reich, Nonexpansive projections and resolvents of accretive operators in Banach spaces, *Houston J. Math.* 3 (1977), 459–470.

[53] Y. Brudnyi and P. Shvartsman, Stability of the Lipschitz extension property under metric transforms, *Geom. Funct. Anal.* 12 (2002), 73–79.

[54] F. Bruhat and J. Tits, Groupes réductifs sur un corps local, *Inst. Hautes Études Sci. Publ. Math.* (1972), 5–251.

[55] S. M. Buckley, K. Falk and D. J. Wraith, Ptolemaic spaces and CAT(0), *Glasg. Math. J.* 51 (2009), 301–314.

[56] D. Burago, Y. Burago and S. Ivanov, *A course in metric geometry*, Graduate Studies in Mathematics 33, American Mathematical Society, Providence, RI, 2001.

[57] H. Busemann, Spaces with non-positive curvature, *Acta Math.* 80 (1948), 259–310.

[58] H. Busemann, *The geometry of geodesics*, Academic Press Inc., New York, N. Y., 1955.

[59] C. H. Chen, Warped products of metric spaces of curvature bounded from above, *Trans. Amer. Math. Soc.* 351 (1999), 4727–4740.

[60] P. R. Chernoff, Note on product formulas for operator semigroups, *J. Functional Analysis* 2 (1968), 238–242.

[61] P. R. Chernoff, *Product formulas, nonlinear semigroups, and addition of unbounded operators*, American Mathematical Society, Providence, R. I., 1974, Memoirs of the American Mathematical Society, No. 140.

[62] T. Christiansen and K. T. Sturm, Expectations and martingales in metric spaces, *Stochastics* 80 (2008), 1–17.

[63] B. Clarke, Geodesics, distance, and the $\mathrm{CAT}(0)$ property for the manifold of Riemannian metrics, *Math. Z.* 273 (2013), 55–93.

[64] B. Clarke and Y. Rubinstein, Ricci flow and the metric completion of the space of Kähler metrics, *Amer. J. Math.* 135 (2013), 1477–1505.

[65] B. Clarke and Y. A. Rubinstein, Conformal deformations of the Ebin metric and a generalized Calabi metric on the space of Riemannian metrics, *Ann. Inst. H. Poincaré Anal. Non Linéaire* 30 (2013), 251–274.

[66] P. L. Combettes and J. C. Pesquet, *Proximal splitting methods in signal processing*, Fixed-point algorithms for inverse problems in science and engineering, Springer Optim. Appl. 49, Springer, New York, 2011, pp. 185–212.

[67] P. Combettes, *Fejér monotonicity in convex optimization*, Encyclopedia of Optimization (C. Floudas and P. Pardalos, eds.), Kluwer, Boston, MA, 2001.

[68] M. G. Crandall and T. M. Liggett, Generation of semi-groups of nonlinear transformations on general Banach spaces, *Amer. J. Math.* 93 (1971), 265–298.

[69] M. Culler and K. Vogtmann, Moduli of graphs and automorphisms of free groups, *Invent. Math.* 84 (1986), 91–119.

[70] S. K. Donaldson, *Symmetric spaces, Kähler geometry and Hamiltonian dynamics*, Northern California Symplectic Geometry Seminar, Amer. Math. Soc. Transl. Ser. 2 196, Amer. Math. Soc., Providence, RI, 1999, pp. 13–33.

[71] S. K. Donaldson, *Conjectures in Kähler geometry*, Strings and geometry, Clay Math. Proc. 3, Amer. Math. Soc., Providence, RI, 2004, pp. 71–78.

[72] A. Dress, K. Huber, J. Koolen, V. Moulton and A. Spillner, *Basic phylogenetic combinatorics*, Cambridge University Press, Cambridge, 2012.

[73] R. Durrett, *Probability: theory and examples*, fourth ed, Cambridge Series in Statistical and Probabilistic Mathematics, Cambridge University Press, Cambridge, 2010.

[74] P. B. Eberlein, *Geometry of nonpositively curved manifolds*, Chicago Lectures in Mathematics, University of Chicago Press, Chicago, IL, 1996.

[75] O. Ebner, Nonlinear Markov semigroups and refinement schemes on metric spaces, *preprint, arXiv:1112.6003*.

[76] P. Enflo, Uniform structures and square roots in topological groups. I, *Israel J. Math.* 8 (1970), 230–252, II, *Israel J. Math.* **8** (1970), 253–272.

[77] P. Enflo, On a problem of Smirnov, *Ark. Mat.* 8 (1969), 107–109.

[78] P. Enflo, On the nonexistence of uniform homeomorphisms between L_p-spaces, *Ark. Mat.* 8 (1969), 103–105 (1969).

[79] K. J. Engel and R. Nagel, *A short course on operator semigroups*, Universitext, Springer, New York, 2006.

[80] A. Es-Sahib and H. Heinich, *Barycentre canonique pour un espace métrique à courbure négative*, Séminaire de Probabilités, XXXIII, Lecture Notes in Math. 1709, Springer, Berlin, 1999, pp. 355–370.

[81] R. Espínola and A. Fernández-León, $\mathrm{CAT}(k)$-spaces, weak convergence and fixed points, *J. Math. Anal. Appl.* 353 (2009), 410–427.

[82] M. Fabian, P. Habala, P. Hájek, V. Montesinos and V. Zizler, *Banach space theory*, CMS Books in Mathematics/Ouvrages de Mathématiques de la SMC, Springer, New York, 2011.

[83] A. Feragen, S. Hauberg, M. Nielsen and F. Lauze, Means in spaces of tree-like shapes, *International Conference of Computer Vision (ICCV)* (2011).

[84] A. Feragen, P. Lo, M. de Bruijne, M. Nielsen and F. Lauze, Towards a theory of statistical tree-shape analysis, *IEEE Transactions of Pattern Analysis and Machine Intelligence, in press, arXiv:1207.5371v1*.

[85] O. P. Ferreira and P. R. Oliveira, Proximal point algorithm on Riemannian manifolds, *Optimization* 51 (2002), 257–270.

[86] T. Foertsch, A. Lytchak and V. Schroeder, Nonpositive curvature and the Ptolemy inequality, *Int. Math. Res. Not. IMRN* (2007), Art. ID rnm100, 15.

[87] Z. Frolík, Concerning topological convergence of sets, *Czechoslovak Math. J* 10(85) (1960), 168–180.

[88] T. Gelander, A. Karlsson and G. A. Margulis, Superrigidity, generalized harmonic maps and uniformly convex spaces, *Geom. Funct. Anal.* 17 (2008), 1524–1550.

[89] A. Genel and J. Lindenstrauss, An example concerning fixed points, *Israel J. Math.* 22 (1975), 81–86.

[90] J. R. Giles, *Convex analysis with application in the differentiation of convex functions*, Research Notes in Mathematics 58, Pitman (Advanced Publishing Program), Boston, Mass., 1982.

[91] K. Goebel and W. A. Kirk, *Topics in metric fixed point theory*, Cambridge Studies in Advanced Mathematics 28, Cambridge University Press, Cambridge, 1990.

[92] K. Goebel and S. Reich, *Uniform convexity, hyperbolic geometry, and nonexpansive mappings*, Monographs and Textbooks in Pure and Applied Mathematics 83, Marcel Dekker Inc., New York, 1984.

[93] A. Goldberg and R. Tarjan, A new approach to the maximum-flow problem, *J. Assoc. Comput. Mach.* 35 (1988), 921–940.

[94] P. Grohs, Wolfowitz's theorem and convergence of consensus algorithms in Hadamard spaces, *to appear in Proc. Amer. Math. Soc.*

[95] M. Gromov, *Hyperbolic groups*, Essays in group theory, Math. Sci. Res. Inst. Publ. 8, Springer, New York, 1987, pp. 75–263.

[96] M. Gromov and R. Schoen, Harmonic maps into singular spaces and p-adic superrigidity for lattices in groups of rank one, *Inst. Hautes Études Sci. Publ. Math.* (1992), 165–246.

[97] M. Gromov, *Metric structures for Riemannian and non-Riemannian spaces*, Progress in Mathematics 152, Birkhäuser Boston Inc., Boston, MA, 1999, Based on the 1981 French original [MR0682063 (85e:53051)], With appendices by M. Katz, P. Pansu and S. Semmes, Translated from the French by Sean Michael Bates.

[98] O. Güler, On the convergence of the proximal point algorithm for convex minimization, *SIAM J. Control Optim.* 29 (1991), 403–419.

[99] B. Halpern, Fixed points of nonexpanding maps, *Bull. Amer. Math. Soc.* 73 (1967), 957–961.

[100] H. S. Hundal, An alternating projection that does not converge in norm, *Nonlinear Anal.* 57 (2004), 35–61.

[101] M. M. Israel, Jr. and S. Reich, Extension and selection problems for nonlinear semigroups in Banach spaces, *Math. Japon.* 28 (1983), 1–8.

[102] J. Jost, Equilibrium maps between metric spaces, *Calc. Var. Partial Differential Equations* 2 (1994), 173–204.

[103] J. Jost, Convex functionals and generalized harmonic maps into spaces of nonpositive curvature, *Comment. Math. Helv.* 70 (1995), 659–673.

[104] J. Jost, *Nonpositive curvature: geometric and analytic aspects*, Lectures in Mathematics ETH Zürich, Birkhäuser Verlag, Basel, 1997.

[105] J. Jost, *Nonlinear Dirichlet forms*, New directions in Dirichlet forms, AMS/IP Stud. Adv. Math. 8, Amer. Math. Soc., Providence, RI, 1998, pp. 1–47.

[106] J. Jost, *Riemannian geometry and geometric analysis*, fifth ed, Universitext, Springer-Verlag, Berlin, 2008.

[107] T. Kato, Nonlinear semigroups and evolution equations, *J. Math. Soc. Japan* 19 (1967), 508–520.

[108] T. Kato and K. Masuda, Trotter's product formula for nonlinear semigroups generated by the subdifferentials of convex functionals, *J. Math. Soc. Japan* 30 (1978), 169–178.

[109] W. A. Kirk, Fixed point theorems in $\mathrm{CAT}(0)$ spaces and $\mathbb{R}$-trees, *Fixed Point Theory Appl.* (2004), 309–316.

[110] W. A. Kirk, *Geodesic geometry and fixed point theory. II*, International Conference on Fixed Point Theory and Applications, Yokohama Publ., Yokohama, 2004, pp. 113–142.

[111] W. A. Kirk and B. Panyanak, A concept of convergence in geodesic spaces, *Nonlinear Anal.* 68 (2008), 3689–3696.

[112] W. A. Kirk, *Geodesic geometry and fixed point theory*, Seminar of Mathematical Analysis (Malaga/Seville, 2002/2003), Colecc. Abierta 64, Univ. Sevilla Secr. Publ., Seville, 2003, pp. 195–225.

[113] M. Kirszbraun, Über die zusammenziehende und Lipschitzsche Transformationen, *Fundamenta Math.* (1934), 77–108.

[114] U. Kohlenbach, *Applied proof theory: proof interpretations and their use in mathematics*, Springer Monographs in Mathematics, Springer-Verlag, Berlin, 2008.

[115] U. Kohlenbach and L. Leuştean, Effective metastability of Halpern iterates in $\mathrm{CAT}(0)$ spaces, *Adv. Math.* 231 (2012), 2526–2556.

[116] U. Kohlenbach and L. Leuştean, Addendum to "Effective metastability of Halpern iterates in $\mathrm{CAT}(0)$ spaces" [Adv. Math. 231 (5) (2012) 2526–2556], *Adv. Math.* 250 (2014), 650–651.

[117] U. Kohlenbach and L. Leuştean, Mann iterates of directionally nonexpansive mappings in hyperbolic spaces, *Abstr. Appl. Anal.* (2003), 449–477.

[118] E. Kopecká and S. Reich, A note on the von Neumann alternating projections algorithm, *J. Nonlinear Convex Anal.* 5 (2004), 379–386.

[119] E. Kopecká and S. Reich, Asymptotic behavior of resolvents of coaccretive operators in the Hilbert ball, *Nonlinear Anal.* 70 (2009), 3187–3194.

[120] E. Kopecká and S. Reich, A note on alternating projections in Hilbert space, *J. Fixed Point Theory Appl.* 12 (2012), 41–47.

[121] N. J. Korevaar and R. M. Schoen, Sobolev spaces and harmonic maps for metric space targets, *Comm. Anal. Geom.* 1 (1993), 561–659.

[122] L. Kovalev, Lipschitz retraction of finite subsets of Hadamard spaces, *preprint, arXiv:1406.6742*.

[123] M. A. Krasnoselskij, Two remarks on the method of successive approximations, *Uspehi Mat. Nauk (N.S.)* 10 (1955), 123–127.

[124] K. Kuwae and T. Shioya, Variational convergence over metric spaces, *Trans. Amer. Math. Soc.* 360 (2008), 35–75.

[125] S. Lang, *Fundamentals of differential geometry*, Graduate Texts in Mathematics 191, Springer-Verlag, New York, 1999.

[126] U. Lang, B. Pavlović and V. Schroeder, Extensions of Lipschitz maps into Hadamard spaces, *Geom. Funct. Anal.* 10 (2000), 1527–1553.

[127] U. Lang and V. Schroeder, Kirszbraun's theorem and metric spaces of bounded curvature, *Geom. Funct. Anal.* 7 (1997), 535–560.

[128] J. R. Lee and A. Naor, Extending Lipschitz functions via random metric partitions, *Invent. Math.* 160 (2005), 59–95.

[129] L. Leustean, A quadratic rate of asymptotic regularity for CAT(0)-spaces, *J. Math. Anal. Appl.* 325 (2007), 386–399.

[130] C. Li, G. López and V. Martín-Márquez, Monotone vector fields and the proximal point algorithm on Hadamard manifolds, *J. Lond. Math. Soc. (2)* 79 (2009), 663–683.

[131] C. Li, B. S. Mordukhovich, J. Wang and J. C. Yao, Weak sharp minima on Riemannian manifolds, *SIAM J. Optim.* 21 (2011), 1523–1560.

[132] T. C. Lim, Remarks on some fixed point theorems, *Proc. Amer. Math. Soc.* 60 (1976), 179–182 (1977).

[133] P. L. Lions and B. Mercier, Splitting algorithms for the sum of two nonlinear operators, *SIAM J. Numer. Anal.* 16 (1979), 964–979.

[134] J. Lott and C. Villani, Hamilton-Jacobi semigroup on length spaces and applications, *J. Math. Pures Appl. (9)* 88 (2007), 219–229.

[135] T. Mabuchi, Some symplectic geometry on compact Kähler manifolds. I, *Osaka J. Math.* 24 (1987), 227–252.

[136] W. R. Mann, Mean value methods in iteration, *Proc. Amer. Math. Soc.* 4 (1953), 506–510.

[137] B. Martinet, Régularisation d'inéquations variationnelles par approximations successives, *Rev. Française Informat. Recherche Opérationnelle* 4 (1970), 154–158.

[138] E. Matoušková and S. Reich, The Hundal example revisited, *J. Nonlinear Convex Anal.* 4 (2003), 411–427.

[139] U. F. Mayer, Gradient flows on nonpositively curved metric spaces and harmonic maps, *Comm. Anal. Geom.* 6 (1998), 199–253.

[140] M. Mendel and A. Naor, Expanders with respect to Hadamard spaces and random graphs, *to appear in Duke Math. J.*

[141] M. Mendel and A. Naor, Spectral calculus and Lipschitz extension for barycentric metric spaces, *Anal. Geom. Metr. Spaces* 1 (2013), 163–199.

[142] K. Menger, Untersuchungen über allgemeine Metrik, *Math. Ann.* 100 (1928), 75–163.

[143] E. Miller, M. Owen and S. Provan, Averaging metric phylogenetic trees, *preprint, arXiv:1211.7046v1*.

[144] I. Miyadera and S. Ôharu, Approximation of semi-groups of nonlinear operators, *Tôhoku Math. J. (2)* 22 (1970), 24–47.

[145] B. Mordukhovich and N. M. Nam, Applications of variational analysis to a generalized Fermat-Torricelli problem, *J. Optim. Theory Appl.* 148 (2011), 431–454.

[146] B. S. Mordukhovich, N. M. Nam and J. Salinas, Applications of variational analysis to a generalized Heron problem, *Appl. Anal.* 91 (2012), 1915–1942.

[147] B. S. Mordukhovich, N. M. Nam and J. Salinas, Jr., Solving a generalized Heron problem by means of convex analysis, *Amer. Math. Monthly* 119 (2012), 87–99.

[148] U. Mosco, Convergence of convex sets and of solutions of variational inequalities, *Advances in Math.* 3 (1969), 510–585.

[149] A. Naor and S. Sheffield, Absolutely minimal Lipschitz extension of tree-valued mappings, *Math. Ann.* 354 (2012), 1049–1078.

[150] A. Navas, An L^1 ergodic theorem with values in a non-positively curved space via a canonical barycenter map, *Ergodic Theory Dynam. Systems* 33 (2013), 609–623.

[151] O. Nevanlinna and S. Reich, Strong convergence of contraction semigroups and of iterative methods for accretive operators in Banach spaces, *Israel J. Math.* 32 (1979), 44–58.

[152] J. Neveu, *Discrete-parameter martingales*, revised ed, North-Holland Publishing Co., Amsterdam, 1975, Translated from the French by T. P. Speed, North-Holland Mathematical Library, Vol. 10.

[153] I. Nikolaev, The tangent cone of an Aleksandrov space of curvature $\leq K$, *Manuscripta Math.* 86 (1995), 137–147.

[154] S. I. Ohta, Markov type of Alexandrov spaces of non-negative curvature, *Mathematika* 55 (2009), 177–189.

[155] S. I. Ohta and M. Pálfia, Discrete-time gradient flows and law of large numbers in Alexandrov spaces, *preprint, arXiv:1402.1629v1*.

[156] Z. Opial, Weak convergence of the sequence of successive approximations for nonexpansive mappings, *Bull. Amer. Math. Soc.* 73 (1967), 591–597.

[157] M. Owen and S. Provan, A fast algorithm for computing geodesic distances in tree space, *IEEE/ACM Trans. Computational Biology and Bioinformatics* 8 (2011), 2–13.

[158] M. Owen, Computing geodesic distances in tree space, *SIAM J. Discrete Math.* 25 (2011), 1506–1529.

[159] L. Pachter and B. Sturmfels (eds.), *Algebraic statistics for computational biology*, Cambridge University Press, New York, 2005.

[160] E. A. Papa Quiroz and P. R. Oliveira, Proximal point methods for quasiconvex and convex functions with Bregman distances on Hadamard manifolds, *J. Convex Anal.* 16 (2009), 49–69.

[161] C. Papadimitriou and K. Steiglitz, *Combinatorial optimization: algorithms and complexity*, Prentice-Hall Inc., Englewood Cliffs, N.J., 1982.

[162] A. Papadopoulos, *Metric spaces, convexity and nonpositive curvature*, IRMA Lectures in Mathematics and Theoretical Physics 6, European Mathematical Society (EMS), Zürich, 2005.

[163] G. B. Passty, Ergodic convergence to a zero of the sum of monotone operators in Hilbert space, *J. Math. Anal. Appl.* 72 (1979), 383–390.

[164] A. Pazy, *Semigroups of linear operators and applications to partial differential equations*, Applied Mathematical Sciences 44, Springer-Verlag, New York, 1983.

[165] X. Pennec, P. Fillard and N. Ayache, A Riemannian framework for tensor computing, *International Journal of Computer Vision* 66 (2006), 41–66.

[166] R. R. Phelps, Convex sets and nearest points, *Proc. Amer. Math. Soc.* 8 (1957), 790–797.

[167] S. Reich, Weak convergence theorems for nonexpansive mappings in Banach spaces, *J. Math. Anal. Appl.* 67 (1979), 274–276.
[168] S. Reich, Product formulas, nonlinear semigroups, and accretive operators, *J. Funct. Anal.* 36 (1980), 147–168.
[169] S. Reich, A complement to Trotter's product formula for nonlinear semigroups generated by the subdifferentials of convex functionals, *Proc. Japan Acad. Ser. A Math. Sci.* 58 (1982), 193–195.
[170] S. Reich, Solutions of two problems of H. Brézis, *J. Math. Anal. Appl.* 95 (1983), 243–250.
[171] S. Reich, The asymptotic behavior of a class of nonlinear semigroups in the Hilbert ball, *J. Math. Anal. Appl.* 157 (1991), 237–242.
[172] S. Reich, The alternating algorithm of von Neumann in the Hilbert ball, *Dynam. Systems Appl.* 2 (1993), 21–25.
[173] S. Reich and I. Shafrir, Nonexpansive iterations in hyperbolic spaces, *Nonlinear Anal.* 15 (1990), 537–558.
[174] S. Reich and L. Shemen, A note on Halpern's algorithm in the Hilbert ball, *J. Nonlinear Convex Anal.* 14 (2013), 853–862.
[175] S. Reich and D. Shoikhet, Semigroups and generators on convex domains with the hyperbolic metric, *Atti Accad. Naz. Lincei Cl. Sci. Fis. Mat. Natur. Rend. Lincei (9) Mat. Appl.* 8 (1997), 231–250.
[176] S. Reich and S. Simons, Fenchel duality, Fitzpatrick functions and the Kirszbraun-Valentine extension theorem, *Proc. Amer. Math. Soc.* 133 (2005), 2657–2660 (electronic).
[177] S. Reich and A. J. Zaslavski, Infinite products of resolvents of accretive operators, *Topol. Methods Nonlinear Anal.* 15 (2000), 153–168, Dedicated to Juliusz Schauder, 1899–1943.
[178] S. Reich and A. J. Zaslavski, *Genericity in nonlinear analysis*, Developments in Mathematics 34, Springer, New York, 2014.
[179] J. G. Rešetnjak, Non-expansive maps in a space of curvature no greater than K, *Sibirsk. Mat. Ž.* 9 (1968), 918–927.
[180] R. T. Rockafellar, Monotone operators and the proximal point algorithm, *SIAM J. Control Optimization* 14 (1976), 877–898.
[181] S. Saejung, Halpern's iteration in CAT(0) spaces, *Fixed Point Theory Appl.* (2010), Art. ID 471781, 13.
[182] T. Sato, An alternative proof of Berg and Nikolaev's characterization of $\mathrm{CAT}(0)$-spaces via quadrilateral inequality, *Arch. Math. (Basel)* 93 (2009), 487–490.
[183] S. Semmes, Complex Monge-Ampère and symplectic manifolds, *Amer. J. Math.* 114 (1992), 495–550.
[184] C. Semple and M. Steel, *Phylogenetics*, Oxford Lecture Series in Mathematics and its Applications 24, Oxford University Press, Oxford, 2003.
[185] E. Seneta, *Non-negative matrices and Markov chains*, Springer Series in Statistics, Springer, New York, 2006, Revised reprint of the second (1981) edn. [Springer-Verlag, New York; MR0719544].
[186] I. Shafrir, Coaccretive operators and firmly nonexpansive mappings in the Hilbert ball, *Nonlinear Anal.* 18 (1992), 637–648.
[187] I. Shafrir, Theorems of ergodic type for ρ-nonexpansive mappings in the Hilbert ball, *Ann. Mat. Pura Appl. (4)* 163 (1993), 313–327.
[188] E. N. Sosov, On analogues of weak convergence in a special metric space, *Izv. Vyssh. Uchebn. Zaved. Mat.* (2004), 84–89.
[189] I. Stojkovic, Approximation for convex functionals on non-positively curved spaces and the Trotter-Kato product formula, *Adv. Calc. Var.* 5 (2012), 77–126.

[190] J. Streets, The consistency and convergence of K-energy minimizing movements, *preprint. arXiv:1301.3948v1.*

[191] J. Streets, Long time existence of minimizing movement solutions of Calabi flow, *preprint. arXiv:1208.2718v3.*

[192] K. T. Sturm, Nonlinear Markov operators associated with symmetric Markov kernels and energy minimizing maps between singular spaces, *Calc. Var. Partial Differential Equations* 12 (2001), 317–357.

[193] K. T. Sturm, Nonlinear Markov operators, discrete heat flow, and harmonic maps between singular spaces, *Potential Anal.* 16 (2002), 305–340.

[194] K. T. Sturm, Nonlinear martingale theory for processes with values in metric spaces of non-positive curvature, *Ann. Probab.* 30 (2002), 1195–1222.

[195] K. T. Sturm, *Probability measures on metric spaces of nonpositive curvature*, Heat kernels and analysis on manifolds, graphs, and metric spaces (Paris, 2002), Contemp. Math. 338, Amer. Math. Soc., Providence, RI, 2003, pp. 357–390.

[196] K. T. Sturm, A semigroup approach to harmonic maps, *Potential Anal.* 23 (2005), 225–277.

[197] H. F. Trotter, Approximation of semi-groups of operators, *Pacific J. Math.* 8 (1958), 887–919.

[198] H. F. Trotter, On the product of semi-groups of operators, *Proc. Amer. Math. Soc.* 10 (1959), 545–551.

[199] M. Tsukada, Convergence of best approximations in a smooth Banach space, *J. Approx. Theory* 40 (1984), 301–309.

[200] F. A. Valentine, On the extension of a vector function so as to preserve a Lipschitz condition, *Bull. Amer. Math. Soc.* 49 (1943), 100–108.

[201] K. Vogtmann, What is... outer space?, *Notices Amer. Math. Soc.* 55 (2008), 784–786.

[202] H. Wang and J. S. Marron, Object oriented data analysis: sets of trees, *Ann. Statist.* 35 (2007), 1849–1873.

[203] R. A. Wijsman, Convergence of sequences of convex sets, cones and functions. II, *Trans. Amer. Math. Soc.* 123 (1966), 32–45.

[204] D. Williams, *Probability with martingales*, Cambridge Mathematical Textbooks, Cambridge University Press, Cambridge, 1991.

[205] R. Wittmann, Approximation of fixed points of nonexpansive mappings, *Arch. Math. (Basel)* 58 (1992), 486–491.

[206] J. Wolfowitz, Products of indecomposable, aperiodic, stochastic matrices, *Proc. Amer. Math. Soc.* 14 (1963), 733–737.

[207] S. A. Wolpert, *Geometry of the Weil-Petersson completion of Teichmüller space*, Surveys in differential geometry, Vol. VIII (Boston, MA, 2002), Surv. Differ. Geom., VIII, Int. Press, Somerville, MA, 2003, pp. 357–393.

[208] S. Yamada, On the geometry of Weil-Petersson completion of Teichmüller spaces, *Math. Res. Lett.* 11 (2004), 327–344.

[209] T. Yokota, Convex functions and barycenter on CAT(1)-spaces of small radii, *preprint, http://www.kurims.kyoto-u.ac.jp/~takumiy/.*

[210] H. Yu, Some proof details for asynchronous stochastic approximation algorithms, *Unpublished manuscript* (2012).

[211] A. J. Zaslavski, Inexact proximal point methods in metric spaces, *Set-Valued Var. Anal.* 19 (2011), 589–608.

Index

T

U

V

W

De Gruyter Series in Nonlinear Analysis and Applications

Volume 21
Moshe Marcus, Laurent Véron
Nonlinear Second Order Elliptic Equations Involving Measures, 2013
ISBN 978-3-11-030515-9, e-ISBN 978-3-11-030531-9, Set-ISBN 978-3-11-030532-6

Volume 20
John R. Graef, Johnny Henderson, Abdelghani Ouahab
Impulsive Differential Inclusions: A Fixed Point Approach, 2013
ISBN 978-3-11-029361-6, e-ISBN 978-3-11-029531-3, Set-ISBN 978-3-11-029532-0

Volume 19
Petr Hájek, Michal Johanis
Higher Smoothness in Banach Spaces, 2014
ISBN 978-3-11-025898-1, e-ISBN 978-3-11-025899-8, Set-ISBN 978-3-11-220385-9

Volume 18
Smaïl Djebali, Lech Górniewicz, Abdelghani Ouahab
Solution Sets for Differential Equations and Inclusions, 2012
ISBN 978-3-11-029344-9, e-ISBN 978-3-11-029356-2, Set-ISBN 978-3-11-029357-9

Volume 17
Jürgen Appell, Józef Banas, Nelson José Merentes Díaz
Bounded Variation and Around, 2013
ISBN 978-3-11-026507-1, e-ISBN 978-3-11-026511-8, Set-ISBN 978-3-11-026624-5

Volume 16
Martin Väth
Topological Analysis: From the Basics to the Triple Degree for Nonlinear Fredholm Inclusions, 2012
ISBN 978-3-11-027722-7, e-ISBN 978-3-11-027733-3, Set-ISBN 978-3-11-027734-0